D0892032

About Island Press

Since 1984, the nonprofit organization Island Press has been stimulating, shaping, and communicating ideas that are essential for solving environmental problems worldwide. With more than 1,000 titles in print and some 30 new releases each year, we are the nation's leading publisher on environmental issues. We identify innovative thinkers and emerging trends in the environmental field. We work with world-renowned experts and authors to develop cross-disciplinary solutions to environmental challenges.

Island Press designs and executes educational campaigns, in conjunction with our authors, to communicate their critical messages in print, in person, and online using the latest technologies, innovative programs, and the media. Our goal is to reach targeted audiences—scientists, policy makers, environmental advocates, urban planners, the media, and concerned citizens—with information that can be used to create the framework for long-term ecological health and human well-being.

Island Press gratefully acknowledges major support from The Bobolink Foundation, Caldera Foundation, The Curtis and Edith Munson Foundation, The Forrest C. and Frances H. Lattner Foundation, The JPB Foundation, The Kresge Foundation, The Summit Charitable Foundation, Inc., and many other generous organizations and individuals.

Generous support for this publication was provided by Deborah Wiley.

Swamplands

Swamplands

Tundra Beavers, Quaking Bogs, and the Improbable World of Peat

Edward Struzik

ISLANDPRESS | Washington | Covelo

Library of Congress Control Number: 2021936542

All Island Press books are printed on environmentally responsible materials.

Cover images courtesy of Shutterstock: owl: vector_ann; berries: Cat_arch_angel; globeflower: Elizaveta Melentyeva; orchid: Alina Briazgunova; moth (top): Maryna S.; moth (bottom): Maria Stezhko; beaver: Kovaleva Galina; background: David M. Schrader

Manufactured in the United States of America
10 9 8 7 6 5 4 3 2 1

Keywords: Alaka'i Swamp, Albemarle Peninsula, Arctic, Ash Meadows, Aweme borer, beaver, bog, botany, canal, Central Park, climate change, colonists, cypress, Death Valley, grizzly bear, Everglades, fen, flooding, French drain, Georgian Bay, Great Dismal Swamp, Hudson Bay Lowlands, hydrology, invasive species, maroon, marsh, miasma, mining, moss, nuclear bomb, Olmsted, orchids, peatland, permafrost, pocosin, polar bear, Powhatan, pupfish, rattlesnake, red wolf, restoration, sphagnum, Strathcona Fiord, Tolland Man, tundra, US Fish and Wildlife Service, water retention, wildfire

Contents

Preface

And what sort of a river was it? Was it like an Irish stream, winding through the brown bogs, where the wild ducks squatter up from among the white water-lilies, and the curlews flit to and fro, crying "Tullie-wheep, mind your sheep"; and Dennis tells you strange stories of the Peishtamore, the great bogy-snake which lies in the black peat pools, among the old pine-stems, and puts his head out at night to snap at the cattle as they come down to drink?
— Charles Kinsley, *The Water Babies*[1]

The initial idea for a book on peatlands came to me incrementally during several trips I made to Banks Island, one of 36,563 islands in the Canadian Arctic Archipelago. The first, in 1991, was a month-long paddle down the Thomsen River, which flows gently north into McClure Strait, part of the Northwest Passage that stays frozen for most of the year.

So much of what my wife Julia and I saw on that journey was a composite of contradistinctions that confused and conspired against everything one assumes about the polar world: the familiar blizzard that greeted us on the first day, and the not-so-familiar thunderstorm that arrived with a bang close to the end; barren, rock-hard tundra lying next to butter-soft, spongey peat that gave rise to glorious clumps of mosses, matchstick-sized saxifrage, and pretty white avens; a stark, polar desert river fed by tea-colored streams flowing out of verdant cotton grass meadows where sandhill cranes nest and more than 80,000 muskoxen roam. (Two-thirds of all the muskoxen in the world resided on Banks at the time of our trip.)

The abundance of life on this landscape was baffling in its ability to transcend expectations and simple explanations. Inexplicably, there are six fish species in the Thomsen, as well as 97 mosses and 83 specimens of lichen growing along its banks and in upland meadows.[2] Why, I wondered, while participating in a scientific survey of raptors many years after that canoe trip, do peregrine falcons, gyrfalcons, and rough-legged hawks come to nest here in such relatively large numbers? Is it for the same reason that a half-million snow geese make the trip from California, New Mexico, and Mexico to nest on the west side of the island where I tried and failed to be the first to kayak the complete course of the Bernard River? And how did the wolves on Banks Island evolve to become genetically distinct from other wolves in the Arctic, as Lindsey Carmichael, a former writing student of mine, reported in her award-winning PhD thesis on these ghostly white animals that visited at night, sometimes with a bark, other times with phantom notice of pee sprayed on the side of tent? We were not being welcomed.

Part of the answer, I learned on another, more recent, month-long trip to the island, has to do with deglaciation and flooding, which are key to the accumulation of peat in most parts of the world. Geologists once thought that Banks Island was spared the deep scarring of the last glacial advance—a refugia of sorts that never went into a centuries-long deep freeze where piles upon piles of snow built up and compressed to form thick sheets of ice. But John England, the scientist I spent time with on Banks, is convinced that the island was thinly covered in ice toward the end of the last Ice Age.

When that thin veil of ice pulled back and slowly wasted away, it left behind hundreds of tundra ponds, polygons, and stream beds where water flowed and where swaths of mosses, sedges, dwarf willow, and other Arctic plants grew and slowly decayed into thick layers of peat. Decomposition could not keep up with the living things that grew so well in a part of the Arctic that is warmed, at times, by a massive wind-driven gyre that regulates climate and sea ice formation.

There are other places in the archipelago where there are verdant peatland meadows. The Kuujjua and Nanook river valleys on Victoria Island (southeast of Banks), which I have canoed through; Sverdrup Pass, the Fosheim Peninsula, and Strathcona Fiord on Ellesmere Island in the most northerly parts of the Arctic are others I have hiked and excavated with paleontologists. The approach to Bylot Island by sailboat in the eastern Arctic was memorable for so many geese nesting among predatory Arctic foxes and red foxes that recently arrived from the mainland. But one has to go much farther south to see an expanse of peat as vast as this in the High Arctic and as many birds, plants, and animals as there are here.

As much as I thought of writing a book on peat following my explorations of Banks Island, there was much more to be learned before I got up the gumption to begin putting pen to paper. These peaty landscapes in the circumpolar Arctic, the boreal forest, the Atlantic cedar swamps, mountains, and deserts, and the tropical and temperate forest of the world do not lend themselves to more conventional descriptions quite like alpenglow lighting up mountain vistas, the ripple and splashes of water echoing through grand canyons, lightning flickering across big prairie skies, and cacti flowering along ephemeral streams that wind their way through the sunburnt cliffs of deserts.

Peatlands are difficult to assimilate because they are ambiguous, elusive, dangerous, sublime, and—as a Red River settler once said—"deceitful." Not land or water, but water and land sharing dominance, like the marginal world of a tidal zone, albeit more slowly, less predictably, and sometimes violently.

The eureka moment came on a 66-day mostly solo kayak trip I did from Virginia Falls on the Nahanni River through some of the 82 peatlands along the Liard and the Mackenzie, the second longest river on the continent.[3] I was three-quarters of the way along the Mackenzie when my water filter failed me. The constant rain and snow had passed by then, mercifully, but the enervating heat of the 22-hour long Arctic days was withering. The river water was so silty that letting it sit overnight was not sufficient to provide me with anything close to a clean drink in the morning, or for the remainder of the day. With no trees to shade me, I was badly burned by the sun and so dehydrated that I was on the verge of hallucinating when the river was about to narrow from more than a mile wide to no more than 100 yards wide where it squeezes through a rampart.

And, just then, I saw a stream flowing down a hill, sparkling in the sunlight. The adrenaline returned and I came back to my senses.

The water turned out to be wonderfully clear and cold, but curiously sweet and smelling of smoke. The stream led me up that hill into a vast, endless wetland that had recently been burned along the edges by wildfire. The peaty flavor and smokiness of that water, I realized, was like the flask of peaty whisky I had brought along and had long ago dispatched. I recognized some of the plants, like Labrador tea, blueberry, and the carnivorous sundew. There were many others, like a very-late-blooming Dragon's mouth orchid (*Arethusa bulbosa*—or possibly a more common Calypso) that didn't seem to belong this far north.

What little I knew about bryology—the study of ancient rootless plants such as mosses, liverworts, and hornworts that colonize nutrient-poor soils—was not enough for me to make sense of their weirdness and intrinsic beauty, other than the fact that these miniature forests were dazzling.

What enchanted me as well was a pale green, almost white carpet of reindeer moss—a lichen, in fact—that had survived the burn. Fresh caribou tracks suggested that I may have frightened them off. There were moose here, boreal birds, and, as I learned later, more than 1 percent of Canada's Pacific loons, scaups, and scoters nesting each spring. I saw more signs of life in those four or five hours of sloshing through peat and crunching across lichen than I had seen in the previous fifty days.

When I returned and set up camp along that stream and cooked up a packet of Kraft macaroni and cheese dinner—a very occasional indulgence that had helped sustain me through impoverished college years—I thought I was in heaven. Just as I was about to feast beneath the cool light of the midnight sun, a conspiracy of ravens gurgled, croaked, and krawed overhead. I knew that something was up, and I soon found out what it was when, with my binoculars, I spied a grizzly bear coming down that hill—lured, no doubt, by the smell of my dinner. I closed my tired eyes, took one big mouthful of the salty, cheesy macaroni, folded up the stove, packed my gear, and jumped into the kayak, not happy about being hungry, but certain that I had an introductory story to tell.

I needed to see more of this in order to put a book together, of course.

More than thirty journeys took me through fens, bogs, swamps, salt and freshwater marshes in the Arctic, the foothills and highest elevations of the Rocky Mountains, below sea level in the Mojave Desert, the boreal and temperate forests, the Great Dismal Swamp and the

Pocosin Lakes National Wildlife Refuge, the west coast of Greenland, the foggy blanket bogs of Labrador and Newfoundland, and the tropical bogs I visited on the Hawaiian island of Kauai. It was instructive to see how they compared and contrasted to the peatlands I had previously explored in Scotland, England, Estonia, Finland, Holland, and Lake Baikal in the southern part of east Siberia and Chukotka in the Russian northeast. There were many similarities—and just as many differences.

Peatlands are so much to so many people—*bofedale*, bog, boglach, fen, glade, holm, marsh, mire, moor, muskeg, morass, polder, quagmire, slough, swale, and swamp. They can be dismal, dark, eerie, magical, and enchanting at the same time. They burble and smell, and light up with will-o'-the-wisps, Cajun fairies, and fireflies. They are clumsy (drunken forests) and murderous in sucking up anything that falls into them accidentally—or is thrown in as a human sacrifice. Peatlands are deceitful in that a bog can turn into a fen, and a rapidly thawing fen that is frozen for most of the year in permafrost can turn into a lake and drown a forest.

On occasion, huge chunks of peat rise up from a fen or bog and drift easily and sometimes dangerously, as the Roman writer Pliny the Younger observed when he described islands of peat with sheep on top of them drifting in the wetlands known as Laghetto di Bassano on the bank of the River Tiber in Italy.

Visiting these peatlands alongside ornithologists, entomologists, botanists, wildlife biologists, bryologists, hydrologists, geochemists, orchid hunters, paleontologists, Inuit and First Nations people, and in one case a physicist who was schooled in the phenomenological aspects of quantum gravity, I realized how much I didn't know about peat and how much more there was to discover. Every one of them was a character—which you have to be, to be passionate about bogs and fens, swamps and peaty marshes. Each had a story to tell worthy of a chapter in a book. One who stands out for tenacity was the moth hunter I spent a day with. In his search for a moth that was once thought to be extinct before a living specimen was discovered in 2009, he drove more than 900 miles and spent 123 nights searching for it in wet, buggy peatlands from the Upper Peninsula of Michigan to eastern Saskatchewan.

He finally found it in a fen.

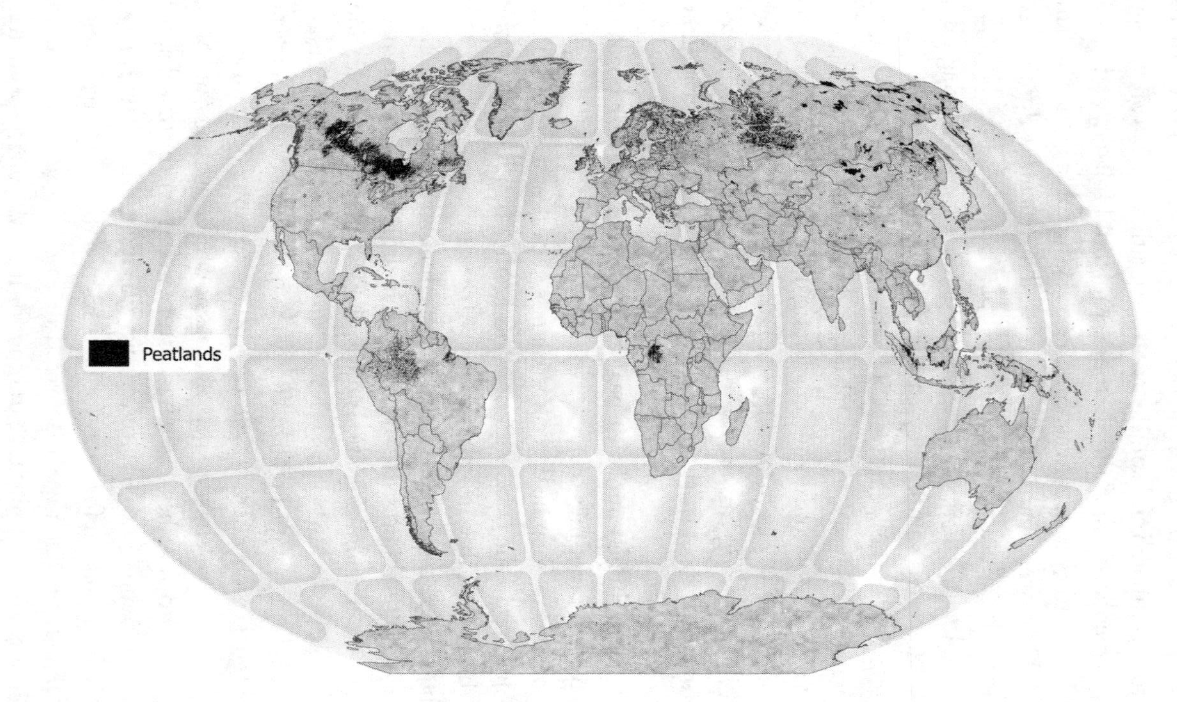

F. M. Southee, K. Richardson, L. Poley, and J. Ray, Wildlife Conservation Society Canada, 2020, https://storymaps.arcgis.com/stories/19d24f59487b46f6a011dba140eddbe7.

Introduction

Sphagnum moss remembers. It recalls
the touchdown of each lark that tumbles
down upon its surface, the slightness of that weight
recorded in the tendrils of each stem. It anticipates
the appetites of flock which graze
upon that wasteland when the rare haze
of summer-heat crisps heather.
The constant tide and toll of weather.
Snow concealing peat and turf like surf,
rolling in with weight of dark clouds curving
around the bleak horizon. The persistent smidge of rain
blurring the land's muted shades year upon damp year again.
— Scottish poet Donald S. Murray[1]

Standing on a mountaintop south of the Arctic Circle in northwestern Canada, I was watching thunderclouds beginning to form over the valley below. The evening sun was still with me as the brooding darkness began to spread across the Nahanni, a 350-mile-long river that flows through three of the deepest and most remote canyons in Canada, and over one of the most spectacular waterfalls on the continent. When the warm, moist air rose up from the river valley and bumped up against the cold mountainside plateau where we were camped, the water vapor in that air mass began to cool, releasing its heat and condensing into billowing clouds that spread out in our direction. Everything very quickly got dark and shadowy.

I could imagine just then how the Mountain Dene, the indigenous people who have lived in this part of the world for thousands of years, came to believe in Ndambadezha, a spirit that inhabits the hotspring vents that rise up through giant, beehive-shaped tufa mounds along the Rabbitkettle River nearby. Ndambadezha was a messenger from the Creator who was sent to Earth to put an end to the chaos that badly behaved humans had inflicted on the natural order and to ward off giant beavers that threatened good hunters paddling along the river.

Had my companions and I been anywhere other than on the edge of a glacier spilling out from an icefield half the size of Manhattan, the vaporous mushrooming might not have seemed as violent as it was that evening. But I figured the temperature was about 15 degrees Celsius colder (27°F colder) where we were than it was in the valley, which is not unusual. The Nahanni is a river I have paddled many times. It can be one of the coldest places in Canada in winter, and one of hottest places in Canada in summer. This warm, increasingly unstable air mass wasn't being nudged upward, as typically happens in the early stages of a thunderstorm. It was more like an active volcano venting aerosols before an eruption.

I was no more than fifty feet from our tent camp when thunder clapped and flashes of lightning electrified the air. I almost waited too long to take refuge from the torrent of rain and hail that followed. My tent turned out to be no match for the blast of wind that came with the moisture. I pressed my feet up against the bending windward wall to keep everything from collapsing, and I flinched every time lightning struck the ground nearby.

When the storm finally passed, the five of us emerged from our shelters to see if our gear was still in place. We canceled our planned evening hike to check on some scientific instruments and weather gauges when a grizzly bear came sauntering across the ice toward the camp. The specter of another storm moving in from across the icefields made it official, soon enveloping us in bone-chilling fog and sopping wet snow. We stayed put for the evening, wondering whether the helicopter would be able to pick us up the next day as planned.

The Bologna and Butterfly Glaciers slide out of an icefield in the Ragged Range along the Yukon and Northwest Territories. (Butterfly was named by glaciologist Mike Demuth in honor of my being tasked with catching butterflies for scientific purposes on the first expedition, a decade earlier.) Thousands of years of snowfall have been compressed into meters-thick sheets of ice that persists because the mountain range intercepts warm Pacific moisture before turning it into snow that falls year-round. The amount of snowfall, however, is no longer enough to prevent the icefield from slowly wasting away as the climate in the western Arctic warms faster than in any other place on earth.

This trip to this icefield was the first of the more than thirty trips I undertook to learn more about peat and its relationship to deglaciation and flooding, which are often a prerequisite to the formation of peat, as well as the many ways in which peat presents itself in culture and the environment, and in global and regional economies.

Peat is partially decomposed plant material that builds up over decades, centuries, and millennia in oxygen-starved, waterlogged conditions where decay can't keep up with growth. The pools of water in the depressions left behind by the scouring of rapidly retreating glaciers are where most but not all peatlands got their start.

Fens, bogs, mangroves, and to a lesser extent, marshes and swamps that accumulate peat cover only a small fraction of the Earth's surface; 3–4 percent is the estimate, but more peatlands are being discovered. Most are located in northern boreal regions of Alaska, northern Canada, Scandinavia, and Russia, where layers of peat can be thirty feet thick or more. But peatlands are also found in temperate forests such as the Tongass in southeast Alaska, the tropical forests of Indonesia, the Australian and European Alps, the west coast of Tasmania, Tierra del Fuego in Patagonia, Hawaii, the Florida Everglades, the Great Dismal Swamp in Virginia and North Carolina, 14,000 feet high in the mountains of Colorado, and below sea level in the Mojave Desert. Reports of the world's largest tropical peatland, in the Republic of Congo, didn't surface until 2017.

Sun-baked Israel and Greece had enormous peatlands in the Hula Valley and the intermontane Drama basin before they were systematically drained. New York state had the Drowned Lands before farmers and timbermen came in and degraded them while driving out the last of the Mohicans. In southern California, wealthy sport hunters dammed the tidal flows into the Bolsa Chica peatlands to make the duck ponds bigger for visiting hunters such as Teddy Roosevelt, the Prince of Wales, and King Gustaf of Sweden.

There are many different kinds of peat. In the Everglades, the building block of peat is sawgrass, a razor-sharp sedge which can grow to heights of seven feet. In salt marshes like those in the Great Bay estuary of New Jersey and the Musquash Estuary in eastern Canada, peat grows from the decay of smooth cordgrass and other salt-tolerant plants. In the Andes, vascular plants form peat. And in most of the tropical peatlands of Africa and southeast Asia, it's palm.

Most of the peatlands in the northern hemisphere and far southern regions are dominated by mosses such as sphagnum, which Canadian ecohydrologist Mike Waddington calls a "supermoss." It's an exquisite looking sponge that can hold between 15 and 25 percent of its weight in moisture. When it grows and spreads out as a mat over water, it can support the weight of a moose, a bear, or a small forest.

But not always.

"Not to be trusted to walk on" is how a Red River settler described the sphagnum bogs and fens of western Canada to *Globe* newspaper readers in 1869. "A moderate-sized river," the correspondent noted, "loses itself under the deceitful turf."[2]

To most everyone but the avid gardener, peat is no more interesting than dirt or, at best, a soil conditioner—which is unfortunate, because you'll need to venture into a peatland if you want to harvest mushrooms, pick wild blueberries, cranberries, and cloudberries, find extremely rare moths, butterflies, and carnivorous orchids, see tree-climbing turtles and massasauga rattlesnakes looking for a place to hibernate, witness endangered woodland caribou taking refuge from wolves and wildfires, or Sumatran tigers hunting tapirs, or Bornean orangutans shaking Sunda slow lorises out of trees, or hear the song of the aquatic warbler, the rarest and the only internationally threatened passerine found in Europe.

One assumes that novelist Smith Henderson (*Fourth of July Creek*) must have visited one of the mountain fens in the Yaak area of Montana before writing so splendidly about "broken cedar, new lime green ferns, and livid mosses. The child's wet warren."[3] But maybe not, because he makes no mention of biting flies and mosquitoes that are so thick, in peatlands like that one, that they can blacken a pot of chili cooking over a campfire or stove in seconds. (I know because the flies did exactly that while I was canoeing along both the Thelon and Back Rivers through the peatlands of the central Arctic of Canada, where, in 1964, botanists Kjeld Holmen and George Scotter were surprised to find seventeen species of sphagnum, many of them newly discovered in the Arctic, in a week.)[4]

Peat is the filter that separates microbes and contaminants from the water that hundreds of millions of people drink; "the kidneys of the landscape" is how botanist William Niering described them in the 1980s when he challenged the once-prevailing view that fens and bogs were "waterlogged wastelands."[5] Peat is the giant sponge that absorbs moisture when rivers flood and when coastal storm surges extend their reach inland. When wet and healthy, a fen or bog can slow or stop a wildfire in its tracks, as Waddington and post-doctoral fellow Sophie Wilkinson noted in describing how the Horse River Fire (2016), the most costly wildfire in Canadian history, may well have been slowed or stopped had fens not been drained as part of a forestry experiment along the lone highway that leads to Fort McMurray, the oil sands capital of the world.

The Amazon and other rain forests get well-deserved attention for the amount of carbon they store and for the exotic plants that are harvested and sometimes pillaged for their medicinal properties. But peatlands sequester 0.37 gigatonnes of carbon dioxide (CO_2) a year—storing more carbon than all other vegetation types in the world combined.[6]

Indigenous people harvest many medicines from bog plants such as Labrador tea, the threeleaf goldthread (savoyane) and from many of the 601 species of fungi that can be found in the world's peatlands. Fungi protect the roots of plants and trees from parasites in return for the plants' giving up nutrients that are hard to come by in acidic bogs and fens. Peat holds in the moisture that the threads of fungi (mycelium) need to grow in order to spread and link the roots of different plants. Without peat to insulate and moisten these threads, 90 percent of the world's terrestrial plants that have a mutually beneficial relationship with fungi would no longer be able to communicate.

For at least 2,000 years, peat bogs were used to preserve wheels of cheese and churns of butter throughout Europe and the British Isles. Murder victims were tossed into peat bogs because no one would dare to look there for the missing person. Periodically, the partially decomposed bodies of human sacrifices as well as mammoths and other prehistoric animals show up in bogs.

Vikings and newcomers to America smelted iron from bog ore consisting of hydrated iron oxide minerals. Jamestown colonists who settled along the boggy shores of Virginia started doing this in 1608, the year after they arrived. The Puritans who settled the Massachusetts Bay Colony built two smelters near present-day Boston in 1644. In 1761, 800 tons of iron were extracted from the bogs of the Saint-Maurice Valley near Trois-Rivières, Quebec, to help the French who were deeply involved in the Seven Years' War and in short supply of iron. The bog-iron furnaces there blasted for nearly 150 years. In more recent times, the Crayola Company harvested iron from the fens in the Colorado Rockies as the base for the color of "Burnt Sienna."

Peat is as much a cultural artifact as it is the sum of so many dead plants and animals. In Scotland, some brands of peated whisky are imbued with aromatic smoke ("peat reek") from phenolic compounds released by peat fires used to dry malted barley. People in Finland refer to their country as *Suomi*, which likely comes from the word *suomaa* meaning "swampland." Some Finns slather themselves with creams of peat and other ingredients in birch-heated saunas that are almost as numerous as the country's adult population. The Welsh celebrate peatlands with an international bog snorkeling event at Waen Rhydd peat bog near Llanwrtyd Wells.

Nineteenth-century amateur bryologists such as F. E. Tripp had a passion for mosses, describing them affectionately as "mysterious children of darkness and decay." In the book *Independent People*, Halldór Laxness, winner of the Nobel Prize in Literature, writes with rare whimsy about "birds of the moor," having "laid their first eggs, yet not having lost the love in their song. Through the heath there ran limpid little streams and round them there were green hollows for the cow, and then there were rocks where the elves lived."[7] It is a very dark tale, but the Icelandic characters in the book embrace the bogs with pleasure, as a young woman does when she wades "barefoot in the lukewarm mud," the mud spurting up between her bare toes and sucking noisily when she lifts her heel.[8] Laxness underscores the beguiling loveliness of the place, where "the feathery mists over the marshes rose twining up the slopes and lay, like a veil, in innocent modesty about the mountain's waist."[9]

Peatlands, however, do not fare so well in English literature.

"All the infections that the sun sucks up / From bogs, fens, flats, on Prosper fall and make him / By inch-meal a disease!" says the sullen, deformed, and savage Caliban in Shakespeare's romance *The Tempest*.

James Joyce's "The Dead" ends with Gabriel contemplating his mortality as he gazes out the hotel window at snow falling on the peaty Bog of Allen, a fictional place which first came up in Charles Lever's novel *Lord Kilgobbin* (1872), as "flat, sad-coloured, and monotonous, fissured in every direction by channels of dark-tinted water, in which the very fish take the same sad colour."[10]

Thomas Hardy's fictional Egdon Heath in *The Return of the Native* (1878) is a "vast tract of unenclosed wild"; Hardy rejects the concept of nature as a positive influence on humankind in favor of Darwin's view of nature as savage and impersonal, and the bogs are where humans struggle for survival, just as, in J. R. R. Tolkien's Middle Earth, elves and men struggled in the Battle of Dagorlad, where the marshes grew sideways (sphagnum moss, presumably) and eventually swallowed up the dead. Legends of elves and little people are remarkably common in peatlands around the world. There are the Pulayaqat of the tundra in the western Arctic, the Menehune of Hawaii's Alaka'i Swamp, the sweet pixies of the moors of southern England, the more malicious boggarts that live in the north country, and the elves that live in rocks in Iceland.

It is somehow fitting that George Orwell wrote his last and most famous book about a doomsday factory while holed up in a drafty house on the moors of the Inner Hebrides island of Jura, digging peat to heat his stove and presumably drinking peated whisky from a now-famous local distillery. "An extremely un-get-atable place," he called it with rueful affection.

Fens and bogs do not fare well in folk culture outside Iceland and some other Scandinavian places because they were once considered to be portals to unearthly realms where gods, crowned with sword-shape asphodels (maiden hair), floated above the heathered moors. In

Northern Europe they were the domain of moss people, green-colored sprites that rose up at night, begging for bread and breast milk and stealing little children if their requests were denied. Indigenous people sometimes feared swamplands as well. Members of the Tsawwassen First Nation in British Columbia have legends of an underground river that flowed from the Pacific into Burns Bog, where spirits would drag people down into the peat, as some like to think happened in 2006 when a foolhardy 34-year-old mountain biker nearly died as he trespassed through the area and sank waist-deep into the bog before being rescued by a helicopter crew. The strange and paranormal stories associated with the Hockomock or Devil's Swamp in southeastern Massachusetts have their origins with the indigenous Wampanoag, who gave the swamp its name, meaning "a place where spirits dwell."

In England, swamp gases that glow at night were "corpse candles" or "will-o'-the-wisps." Another nighttime apparition was Jack-o'-Lantern—a man who, having either murdered a young boy or sold his soul to the devil in return for a gold coin that he would use to buy beer, was condemned to walk the bogs at night carrying a lamp.

In the dark and melancholy poem "The Cold Earth Slept Below," Percy Shelley presents us with a woman who has died of exposure on a cold winter night after being lured into a bleak, presumably peaty landscape by glowing eyes that resemble a "fen-fire" (sometimes called a "corpse candle").

These ghostly apparitions weren't just a British phenomenon that was introduced to North America. In the Netherlands, the *irrbloss* were the souls of unbaptized children shining lights that tricked people into getting lost in the mire. In Bolivia and Argentina, they are *luz mala*, "evil lights." In West Bengal swamps, *aleya* are ghost lights that mark the spot where a person has died.

In order to purge superstitions such as these from the minds of soldiers, including two colonels who thought the glows they saw might have something to do with turpentine that comes from the resin of longleaf pine trees, George Washington and Thomas Paine stirred up the bottom of a New Jersey swamp while waving a torch over the bubbles that rose to the surface. The flash of light that ignited across the water's surface proved that it was swamp gas (methane) that some saw at night, not corpse candles.

Or was it? Some of the glows that people saw may have been fireflies that light up bogs, fens, and cedar swamps. We know so little about them and many other invertebrates that live in peatland ecosystems that we can't say with certainty whether some are dependent on peaty landscapes or just associated with them. One *Photuris* firefly, known as the "Loopy 5," was thought to be a resident of a few far-flung swamps in Tennessee. They were so rare that scientist Lynn Faust, author of *Fireflies, Glow-worms, and Lightning Bugs*, thought she might have only imagined seeing the five-second flash that ends with a flourish before reappearing elsewhere.[11] When she called James E. Lloyd, professor emeritus at the University of Florida, he assured her that he had seen the flash of the so-called swamp sneezer in a North Carolina peatland. The only other place it has been seen since is in a cedar swamp in Mississippi.

African Americans escaping slavery often took refuge in the Great Dismal Swamp along the border of Virginia and North Carolina, believing the plantation owners were too disinclined to venture into these dangerous, phantom-filled forested peatlands to hunt them down. Some stayed and raised families there.

For the Irish, peat is confirmation of identity, for as far back as the fourteenth century, the cutting of peat ("turf") for fuel in Ireland was part of the back-breaking seasonal work levied by landowners. Many Irish students were required to supply peat to help fuel the stoves in their schoolhouses. Some threw smoldering peat briquettes at British soldiers during the Easter Rising of 1916. Turf cutting in Ireland by so-called boggies reached its peak in 1926, when six million tons were produced.[12]

For yak herders, the draining of the Zoigê Marsh (in Chinese, *Ruoergai*) in the northern Sichuan region was a symbol of Communist excess during the Cultural Revolution; this was a monumental draining initiative that did not seek local consultation or offer any apologies until it ended up drying seventeen lakes and countless fens, and degrading a massive amount of the peat that had stored water when it was needed most and drained it when floods threatened.

For Jews and other political prisoners who were forced to dig peat every day in Nazi- and Soviet-era concentration camps, peat was a symbol of oppression as well as defiance. (The Ontario government

As far back as the fourteenth century, the cutting of peat ("turf") for fuel in Ireland was part of the backbreaking seasonal work levied by landowners. (Photo: Mayo County Library, Maggie Blanck Collection.)

in Canada briefly considered using prisoners and alien enemy labor to harvest peat for fuel during the First World War.[13]) "Peat Bog Soldiers" ("*Die Moorsoldaten*") was a song composed by Jews in 1933 and sung by prisoners as they marched into the mires of Börgermoor to labor each day.

> *Heath and bog are everywhere.*
> *Not a bird sings out to cheer us.*
> *Oaks are standing gaunt and bare.*
> *We are the peat bog soldiers,*
> *Marching with our spades to the moor . . .*[14]

The song went on to become an anthem for resistance fighters in Spain and the labor movement in Europe. African American actor, singer, and activist Paul Robeson recorded a version in 1942. The

French Foreign Legion adapted it for one of its marching songs. The Dubliners, an Irish folk group, recorded a version in 2016.

Peat replaced wood as a source of fuel after most of Europe's forests had been felled. The mining or extraction of peat was once a tangible measure of the prosperity in the Netherlands, just as oil and gas is in modern times for nations like Norway. Despite its small population, the Netherlands was a superpower from around 1500 to the 1850s because of the Dutch ability to power industrial activity with peat, wind, and water, and because they successfully drained bogs and fens for agricultural purposes much more efficiently than other, larger, European countries were able to do. This wealth explains how the Dutch East Company routed the British Royal Navy and displaced the Portuguese in the East Indies, and why the company was able to gamble on English explorer Henry Hudson when they sent him across the Atlantic to find a shortcut to the Orient. Instead of spices from the Far East, they got Manhattan and beaver pelts.

More often than not, however, peat was a measure of poverty.

The more peat there was, according to British scholars, the worse off the country. "Poor old Ireland," as social reformer Charles Kingsley called it, was perceived to be destitute because a seventh of the island was covered in "worthless peat."

"No wonder if a Country famous for laziness as Ireland, is abound with them" (that is, peat bogs and fens), wrote Dublin Society fellow William King in 1685. "There is no lack of peat in the United Kingdom, indeed, in so far as it is mere unprofitable and waste land, there is a very great deal too much."[15] More than two centuries later, amateur geologist Ralph Richardson was heartened by the fact that "civilization and agriculture are nibbling at their (peat) borders, and many a fine green sward was but a few years ago a dark and filthy moss."[16] Richardson saw in peat a source of cheap energy that would replace coal, which was becoming increasingly unaffordable to most people during his time.

In Richardson's day, it was a farmer's Christian duty to drain wetlands because the Bible suggested as much to those inclined to do so. In Genesis 1, the world is a shapeless black hole before God separates light from dark, water from water so that land can appear for men and women to eventually plant seeds on. God's work during those first days of creation was seen by some as setting an example for men and women to seek out moist ground and free themselves from the "mud of the mire," as described in Psalm 40. The Bible tells man to be "fruitful, and multiply, and replenish the earth, and subdue it."

In *Pilgrim's Progress*, John Bunyan's Christian allegory, the path to grace is blocked by a foul mire—"The Slough of Despond." When the hero asks the character Help (an angel?) why it remains undrained, he is told that "this miry slough is such a place as cannot be mended: it is the descent whither scum and filth that attends conviction for sin doth continually run, and therefore it is called the Slough of Despond."[17] For the farmer, the message was clear enough. "The Slough of Despond" was a test of faith, courage, and will. Do God's work. Find a way to drain the mire, and you will prosper.

Immigrants brought this ethos to the New World when a third of the United States was covered in wetlands. The first colonists in Virginia set to work almost immediately, planting tobacco on peatlands that were drained to get to the sandy bottoms below. Before he became president of the United States, George Washington was one of a wealthy group of investors who used enslaved laborers in an attempt to drain the Great Dismal Swamp along the Virginia/North Carolina border in order to create farms that they thought would thrive on suitably drained peaty soils. They failed in this, but they succeeded in cutting down most of the Atlantic white cedars and longleaf pines before others fanned out and carried on cutting down trees in cedar swamps northward all the way to Maine.

Even after 250 years of draining wetlands, there was still a lot of water on the American landscape—so much so that Congress passed the Swamp Acts of 1849, 1850, and 1860 to get rid of them. These laws, along with advances in draining technology, led to the dewatering of the Black Swamp in Ohio, the Sacramento–San Joaquin Delta in California and a remnant peatland ecosystem at Ash Meadows, where there is an oasis of peat in the red-hot Mohave Desert.

Estimates suggest that nearly 200,000 square miles of the world's peatlands, an area slightly larger than the state of California, have been systemically drained and degraded, mined and vacuumed to create everything from iron, insulation, tampons, and diapers, as well as the biomass that continues to drive modern-day power plants like the 60 in Finland that provide 5–7 percent of the country's energy needs.[18,19] In 2020, more than 140,000 Swedes and 82,000 Estonians relied on home heating produced by peat. Peat meets 6 percent of Ireland's energy needs and likely much more than that in the remoter regions of Russia.

Some of the more affluent peat-burning countries, such as Ireland, are committed to reducing their dependence on turf. But poor countries like Burundi and Rwanda are moving in the other direction because there are no economic incentives for them to shift to wind and solar power. In Rwanda, where more than half of the population lives in poverty, the government opened its first peat-fired power plant in 2016, with a long-term plan to have peat represent 20 percent of the country's electrical supply. Eighteen of the biggest bogs in the country are being targeted to meet that goal.[20]

For more than a century, Canadians tried turning peat into fuel for locomotives and other uses before realizing that it was easier and more profitable to dry it out and turn it into insulation and garden conditioners. Wainfleet Bog, which once covered 500,000 acres (202,342 hectares) in southern Ontario, shrank to little more than 3,000 acres (1,200 hectares) over an 80-year period. Restoration began in 2000. But twenty years later, Canada was still harvesting about 1.4 million tonnes of peat from other bogs and fens each year.

Peat was once a wartime asset. It was used to make bandages for wounded soldiers and bedding for cavalry horses in the Napoleonic wars and the Franco-Prussian War with Russia in the nineteenth century, and the Russo-Japanese War in the early twentieth century. More than 70,000 tons of peat were required to bed Napoleon's 47,000-horse cavalry during the Waterloo campaign.[21]

When Europe could no longer supply the United States with peat for packing munitions and manufacturing bombs during the Second World War, Canadians stepped in and sent hundreds of tons of peat from Burns Bog in British Columbia to munitions plants near Las Vegas, where it was used as a catalytic agent to refine magnesium for firebombs.

Peatlands were also used as weapons. When Italy switched sides during the Second World War, forcing the Germans to retreat as the

Allied forces landed just south of Rome at Anzio, the Wehrmacht ordered the destruction of the hydraulic works that Mussolini had put in place to drain the Pontine Marshes. The aim was to destroy fertile agricultural lands that would provide food for the enemy and allow malaria-spreading mosquitoes to come back and infect Allied soldiers.[22]

Bombing and military training ranges were typically located in peatlands like the one in Grafenwöhr, Germany, and another that operates near the Alligator National Wildlife Refuge in North Carolina. The logic in choosing such places was simple: there were few people living in them, and there was nothing there that was considered worth preserving.

A very low point for peatlands came in 1948, when New Yorkers began dumping their garbage into the brackish, peaty marshes of Staten Island, close to where Frederick Olmsted, the architect of Central Park, had owned a farm. Freshkills (*kille* being the Dutch word for riverbed) was the biggest garbage dump in the world before New York mayor Rudi Giuliani worked with others to get it closed in 2001.

Abused as they have been, peatlands ecosystems are extremely resilient, as I learned from Mike Waddington when I spent a couple of days with him in the bogs of Georgian Bay, where, to his surprise, one of his graduate students discovered rattlesnakes denning in peat. You can drain and burn bogs and fens, build roads through them and on top of them, and subject them to extended droughts, as Mother Nature sometimes does, he told me, but it usually takes two or more of these disruptions working in tandem to destroy them.

Even then, "bogs go right on behaving like bogs," as Finnish author Tove Jansson says in *The Summer Book*, one of her famous children's stories. A badly degraded peatland will find a way to come back and haunt those who have done it harm. In Indonesia, the recurring peat fires that create toxic haze affecting tens of millions of people are the direct result of clearing peatland forests for palm oil and pulpwood plantations. Burns Bog in British Columbia lit up in 1977, 1990, 1994, 1996, 2005, 2007, and in 2016, the year Waddington published a report that demonstrated that bogs and fens were becoming increasingly vulnerable to fire because of climate change, peat mining, and other disturbances that have happened in the past. Parts of the Great Lakes are being choked with algal blooms because there is no longer enough peat in the adjacent wetlands to filter out nutrients such as phosphorous and nitrogen that flow in from farms.

Like the Everglades, and parts of Louisiana and coastal California, Holland is literally sinking into the sea because of subsidence caused by peat extraction and water diversion, and because there is not enough spongey peat left to soak up rain and the seawater that is surging farther inland into depressions created by the mined-out mires. Trees that were planted on drained bogs and fens in Finland and the Flow Country of Scotland, often with generous tax incentives, have degraded water quality, wildlife habitat, and plant diversity. Such planting often resulted in lodgepole pines with crooked trunks that could not withstand the wind, or fight off the attacks of the pine beauty moth, which developed a taste for the buds of Scots pine that were rooted in treeless fen country. In 2019, thousands of fish washed up dead on the shores of Rautalami Lake in Finland because of the acidic discharge from a peat production plant.

Bog bursts and peat slides such as the one that came close to killing British novelists Emily and Anne Brontë in 1824 are on the rise in the United Kingdom. In November 2020, a peat slide at Meenbog Wind Farm near Ballybofey in the Republic of Ireland blackened one of the most important Atlantic salmon–spawning rivers in all of the European Union.

Peat-rich Canada and Alaska aren't quite there yet, unless one considers the alarming fallout as frozen peatlands begin to thaw and release carbon stored in permafrost. But peatland ecosystems in other parts of North America continue to be scarified by urban expansion, oil and gas development, coal mining, and hydroelectric projects. Ninety percent of the pre-settlement wetlands in the city of Calgary have been lost to urban development.

The peatland complex that lies within the transmission zone of the $7.5-billion Muskrat Falls hydroelectric project in Newfoundland and Labrador is so vast that Nalcor Energy, the government-owned company that is behind the project, has allowed that "it is not practical to assess the functional performance of individual wetlands." Why? Because there are so many of them.[23]

The Brintnell-Bologna Icefield was a good place to see how peat began to form as the Last Glacial Period petered out and created cool, wet conditions that allowed mosses and other plants to grow. Glaciologist Mike Demuth, the expedition leader of our small group of scientists, had invited me to come along to see how far back the glacier had retreated since we were last there in late winter, when the temperatures dropped to −30 degrees Celsius (−22°F) in the evening.

It was not easy getting there, though. The most economical way was for us was to drive 20 hours from Edmonton in southern Canada to Watson Lake on the Alaska Highway in the Yukon, and then another four hours from there along a narrow, impossibly bumpy, mud-covered mining road that crosses into the Northwest Territories. From that remote mine, we flew in to the icefield by helicopter.

Seeing the glacier this third time in a decade was like looking at a childhood friend whom one hasn't seen in decades, trying to reconcile the gray hair, wrinkles, and patchy skin with the smooth face, thick head of hair, and youthful bounce we'd seen before. Soot from wildfires, as well as dirt and dust from windstorms and rockfalls, had turned the icy blue surface of the lower part of the glacier into shades of brown and gray. The bones of a caribou and the freeze-dried carcass of a porcupine had melted out of a crevasse that the animals had probably slipped into after a snowstorm had camouflaged the danger. Along the toe of the glacier, torrents of meltwater poured into depressions that had been scoured out during the advance and retreat of ice. Sunlight reflecting off rock flour suspended in the icy water column gave the pools a turquoise hue. (Rock flour is what is left behind when a glacier grinds its way across a rocky landscape.)

It was a scene that many indigenous people saw a lot of 8,000 to 13,000 years ago as the ice sheets began to retreat and vast tracts of land flooded. Nothing much was growing on the shores of the glacial pond where we drew our water. The early colonizers included mossy saxifrage, alpine bluegrass, aquatic sedges, mountain avens, bog birch, Arctic willow, and dwarf fireweed, which is also one of the first plants to rise up after a wildfire.

A half mile down the valley where I was netting butterflies on the first expedition, plants were taller and shrubbier, and the moss cover thicker and spongier. Eight- to twelve-foot-tall black spruce trees had rooted themselves on hummocks where raging streams of meltwater couldn't mow them down. Orange and green lichen were growing slowly on boulders that were constantly clacking as they were being piled up by a spiderweb of glacial streams fanning out in all directions. (The size of a lichen can reveal the timing of glacial retreat.)

Farther downstream toward the Nahanni Valley, where the cirque and valley glaciers have been absent for a century or more, there were white spruce, aspen, birch, lowbush blueberries, crowberries, Labrador tea, leatherleaf, tamarack, and carpets of caribou moss, which is a lichen, not a moss. The glacial transformation ended abruptly along the edges of the river valley, where ice flowing in from the west and east during the Last Glacial Period stopped short and never got a chance to make its mark.

This cold, moist, post-glacial landscape along the edges of the Brintnell-Bologna Icefield represents the kind of conditions in which most of world's peat began to form. Typically, plants that die, leaves that are shed, and insects and animals that perish are broken down by bacteria, insects, and chemicals in the ground. Anyone who composts knows how this works. This process does not happen as quickly in cold, wet regions and on soil that is frozen and covered in snow for most of the year.

As the partially decomposed organic matter piles up over decades, centuries, and millennia, the oxygen that bacteria and insects need to survive becomes increasingly scarce at the lower depths. This partially decomposed fibrous material we call *peat* can be more than 30 feet deep in the Scotty Creek region 190 miles downstream of the Brintnell-Bologna Icefield, 24 feet deep and more in many parts of the British Isles, Scandinavia, and the Columbian Andes, 3–16 feet deep in the Hudson Bay Lowlands and the boreal forest of western Canada, and between 2 and 5 feet deep in places such as Georgian Bay in Ontario.[24]

Shallower anaerobic layers of peat can be found throughout the Rocky Mountains from Colorado into Canada and on Staten Island and Brooklyn's Prospect Park, where the southward advance of ice stopped 12,000–13,000 years ago. A peatland ecosystem is any place where peat is at least about a foot deep. In North America, it's sixteen inches.

I had assumed that all peatland ecosystems were formed in the same way until I talked to Dale Vitt, a Southern Illinois University bryologist who may well be the longest-living peatland ecologist still actively investigating the wondrous and sometimes elusive nature of these ecosystems. Vitt has published well over 300 papers on the subject, but he really won me over when he pointed out that a photo I had taken of a spectacular yellow, green, and red moss I had found along a dried-out stream in a High Arctic desert was a "dung moss." "*Tetraplodon* is the genus," he noted. "In the Arctic it often grows on muskox dung but also on lemming carcasses."

There are several ways in which peatlands form, according to Vitt. All of them involve a form of swamping. Over time, in many cases, the rock powder and other sediments in glacial meltwater settles down to create or add to an impervious layer of clay. As the ice continues to recede, the water levels rise and flood into the surrounding shrubs and forest where drainage is often poor. The grasses, sedges, shrubs and trees in the flooded forest will eventually die off, setting the stage for wetland development and the spread of more-resilient, moisture-loving plants like sphagnum moss.

Mosses such as sphagnum are the master builders of peat in the boreal world. They are, for want of a better term, serial killers. They snuff out vascular plants by taking up positively charged ions such as calcium and ammonium while releasing hydrogen ions that acidify the water and soil. As long as there is sufficient moisture, mosses will grow upward and outward, periodically shooting out capsules that contain hundreds of thousands of spores that are no more than several microns in diameter. The more moss there is present in cool, moist conditions, the better the chance of peat forming.

Mosses also decompose more slowly than vascular plants. Some mosses are almost indestructible. Catherine La Farge, a bryologist I know and spent time with on Banks Island in the Arctic, found moss fragments that had been entombed in a High Arctic glacier for 400 years. Like stem cells, they had the capacity to regrow and form a new plant.[25]

The impressive flooding that continues to flow from the Brintnell-Bologna Icefield is barely a trickle compared to the massive swamping that took place when there were two-miles-thick sheets of rapidly melting ice that stretched from Vancouver, Canada, on the Pacific coast to Staten Island, Prospect Park, and Central Park in New York, and in small pockets in and around high country like Yosemite in the Sierra Nevada, Park County in Colorado, and the Wasatch Range in Utah. As those ice sheets and alpine icefields began melting, they left behind enormous bodies of water like the Great Salt Lake in the West and the Great Lakes in the East, and smaller, shallower ones like kettle lakes, sloughs, marshes, swamps, bogs, and fens throughout the boreal world.

The meltdown wasn't always gradual. Sometime around 13,000 years ago, the natural ice dam that held back meltwater in Glacial Lake Iroquois, which was an expanded version of Lake Ontario, collapsed and sent a biblical flood of water down the Hudson River Valley, past Manhattan, Brooklyn, and Staten island into the North Atlantic. The water in the lake dropped by as much as 400 feet.

The onset of peatlands in western Canada, where nearly a quarter of the landscape is covered in peat, was delayed for several thousand years because it was much drier there than it was in the East. Around 8,000 years ago, the first of three climatic oscillations that brought wetter weather, cycling in every 1,400 years, flooded forests and killed trees that were growing on dry mineral soil. Mosses and other peat-forming plants eventually took over before peat-loving trees such as black spruce, tamarack, pine, birch, and aspen grew up around bogs and fens. Most of Canada's oil sands companies extract bitumen beneath these deep layers of peat.

Most peatland ecosystems evolved after the retreat of the ice sheets during the Last Glacial Period that ended about 12,000 years ago. But in the fall of 2020, Monika Ruwaimana and her colleagues described one in the West Kalimantan province of Indonesia that is about 48,000 years old, with approximately 4 million acres (1.6 million hectares) of peat. What makes it even more extraordinary is that the peat is exceptionally deep, as much as sixty feet in some places.[26]

Peatlands tend to be flat or undulating, so the moisture that is needed to keep them from drying out stays put. Bogs are fed by rain and additionally sustained in some places, like northern Scotland, Newfoundland, and Labrador, by persistent fog and cool weather. Because precipitation is a poor source of ions, bogs have very low concentrations of cations and anions other than hydrogen. In contrast, fens rely on surface and mineral groundwater, which tends to make them less acidic or alkaline. The distinctions are important, as Colorado State peatland ecologist David Cooper told me, because bogs and fens have different hydrologic regimes, occur in different climate regions, support different plants and animals, and are disturbed by different processes. That explains why there are no bogs in the western United States, save one that was recently found by Cooper and his colleagues on the western Olympic Peninsula in Washington state.

There are so many variations of each that it can boggle the mind of anyone who is not a specialist. Fens can range from acid to basic, depending on their source waters. In the South Park region of the Colorado Rockies, limestone produces alkaline water and extremely rich fens, but in other areas, such as parts of Alaska and Canada, poor fens are quite acidic.

One industrious person working for the International Peatland Society made a list of almost 7,000 peat and peatland-related terms. *Palsas* are elevated domes of frozen peat that are found in the Hudson Bay Lowlands, in the alpine country of MacMillan Pass in the Yukon just south of the Brintnell Icefield, and in other regions of the Arctic. Unlike *pingos*, which are much bigger and grow below the active layer of permafrost, palsas rise up within the layer that does not stay frozen year-round. Polar bears in western Hudson Bay migrate many miles inland into the boreal forest of the Hudson Bay Lowlands to dig birthing dens into the side of palsas.

Quaking bogs are actually fens in many cases. Henry David Thoreau often visited a quaking bog near his cabin in the backwoods of Massachusetts. They are thick mats of peat that grow along edges of a lake or large bog. Walking across them, as I did in MacMillan Pass in the Yukon, in a bog near Lake Baikal in Siberia, and in the Lower Saint Lawrence region of eastern Quebec, is akin to standing on one of those first-generation waterbeds that didn't have enough baffles to prevent water from sloshing back and forth.

Floating islands form when parts of those thick mats of peat break away from the shore of a quaking bog. They were long deemed to be so rare and peculiar that *Nature*, the venerable scientific journal, once described them as a "remarkable freak of nature."[27] This view prevailed for more than a century. In 2005, the *New York Times* marveled at how a floating island of peat in Springfield, Massachusetts, made a beeline to a man's shore property during a windstorm, "crushing his three-foot chain-link fence, swamping his red-blue-and-purple flagstone patio, wrecking his dock, flooding his shed, hobbling his weeping willow, and drowning the oregano, cilantro, tomatoes, and peppers in his garden. Then, with an insouciant shrug, it came to a standstill in Mr. Renna's backyard, an interloper squatting in stubborn silence." The *Times* quoted an environmental official who claimed that this was one of only two floating islands in North America.

Floating islands of peat are, in fact, so common that the state of Minnesota issues permits to people who make a part-time living lassoing them before or after they do damage. In Florida, where trees can grow on top of floating island fens, they act as floating filters, taking up dangerously high levels of nutrients that flow in from nearby farms. In Lake Malawi, in Africa, they facilitate the dispersal of fish, possibly increasing gene flow between species that are related but geographically isolated.[28] Some of the bigger floating islands that detach from coastal regions have been spotted by sailors 1,000 miles out to sea.[29]

Blanket bogs are relatively rare but exceptionally showy for their remarkably diversity. They develop in oceanic highland areas like the Flow Country of Scotland and in coastal regions of Alaska, British Columbia, and Newfoundland and Labrador where a great deal of rain falls on gently sloping hills and valleys that tend to be poorly drained, foggy, and cool for a good part of the year. Nearly a quarter of all of Scotland is covered in blanket bogs dominated by 34 species of sphagnum.

Cataract bogs are those that feature a permanent freshwater stream.

String bogs have a varied landscape, with low-lying "islands" interrupting the saturated bog ecosystem.

Valley bogs develop in shallow valleys.

The nomenclature gets increasingly complicated when you consider there are rich fens, poor fens, and those in between. The richer the fen, the more nutrients there are available for plants and animals. An "extremely rich fen" is what David Cooper found at 10,000 feet in the mountains of Central Colorado in the early 1980s. It contains more rare plant species than any other wetland in Colorado.

Cooper's studies of mountain fens have confirmed what should have been obvious long ago to national park officials like those who in Yosemite who once drained a fen to create a parking lot. (It's since been restored). In places such as the Sierra Nevada, where fens represent less than 0.1 percent of the mountain landscape, they are perennially wet refuges for scores of plants, amphibians, and aquatic invertebrates, and in many cases the places where wildfires slow or stop. These can include sloping fens, basin fens, mound fens, and lava fens.[30]

The deeper you dig into the nomenclature, as British author Robert Macfarlane has done so splendidly in his book *Landmarks*, the more you find that there are dozens of words for peatlands that are no longer part of the common vocabulary. In Norfolk, a floating island was called a *bover*. In Cumbria, *watter-sick* was peat that was heavily saturated with moisture. In Orkney, *yarpha* was peat that is full of roots and fibers. In Wales, a bog was called *mign*. More accessible to the uneducated is "MAMBA country"—the peatlands of the Elan Valley of Wales where there are Miles And Miles of Bugger-All.

In Canada, *muskeg* is the word that the Mushkegowuk people of the Hudson Bay Lowlands used to describe fens, bogs, marshes, and pretty much everything that is wet, squishy and buggy in summer and icebox cold in winter. (In the 1950s and '60s, Canada's National Research Council sponsored a number of conferences aimed at dealing with the "muskeg problem.") In Virginia and North Carolina, raised peatlands are called *pocosins*, the Algonquin word for "swamp on a hill." The Mohawks of the Hudson River Valley use the word *masspootupaug* to describe a boggy meadow.

The one I like is *niaquptak*, which my friend Piita Irniq told me is derived from the Inuit word *Niaquq*, meaning head. Piita was born and raised in tents and igloos in the Arctic long before becoming the second commissioner of the Inuit territory of Nunavut when it was established in 1999. "The tufts of grass on these wetlands resemble so many heads," he told me

The thing that peatlands have all had in common is that they were and still are underappreciated. In western Canada, where nearly half of the boreal forest floor is covered in peat, there is just one peatland ecologist working for the government. The situation in Europe is much better, because the public and many politicians now realize that the degradation of so much peatland has been an economic and environmental disaster. Had I tried, I might have been able to trace all the peatland ecologists working in North America to a small group of modern-day mentors that includes Dale Vitt and David Cooper in the United States, Line Rochefort and Nigel Roulet in Quebec, William Shotyk, Kevin Devito, Lee Foote, and Suzanne Bayley in Alberta, and Mike Waddington, Phil Marsh, and Jonathan Price in Ontario. There is a growing number of younger mentors like Merritt Turetsky, the self-described "queenofpeat," from the University of Colorado, Boulder, but not nearly as many as you would find in most fields of environmental study.

Several United Nations World Heritage sites contain peatlands, but none of them were established for that reason. Only recently have the Great Vasyugan Mire of Russia, the Okefenokee National Wildlife Refuge of the United States, and the Céide Fields and North West Mayo Boglands of Ireland been added to tentative lists. Scotland is finally planning to make a case for the Flow Country in 2023.

Many people, including some scientists and public policy decision-makers, continue to underestimate the importance of peatland ecosystems because of the perceived dearth of plant and animal species. This, of course, is true if you compare the peat-rich boreal forest region to the Amazon rain forest. But it is a false slight, given all the virtues of peat. One to three billion birds fly north to the boreal peatlands of North America each spring with a reproductive purpose that results in three to five billion of them migrating back in fall. Some of these birds are migrating from one peatland ecosystem like the Great Dismal Swamp to another peatland like the Hudson Bay Lowlands. The whooping crane, one of the world's most endangered birds, flies north each spring from the spongy shores of the Aransas National Wildlife Refuge in Texas to the fens of Wood Buffalo National Park in northern Canada.

Without peatlands, most of North America's finches and warblers, and 80 percent of the waterfowl, would be forced to find another place to nest. Lemmings and hares would lose an important food source if there were no moss capsules to eat. So too would the Norwegian grouse and the blue and marsh tits of Britain. Moorhens and snow buntings rely on moss to some extent, as do many Arctic birds, especially chicks when they are too small to handle large food items.[31]

Tropical peatlands are in a category of their own. In the early 1990s, when botanists Jack Rieley (University of Nottingham) and Susan Page (University of Leicester) developed an interest in the ecology of tropical peatland ecosystems, the prevailing view among scientists was that the plant and animal communities were sparse and uninteresting.[32] But now we know that Borneo alone is home to 10,000–15,000 species of flowering plants as well as 37 endemic bird species and 44 endemic mammal species. More than 400 terrestrial species are classified as threatened by the International Union of the Conservation of Nature.[33] There are likely a lot more to discover.

The fact that we have gone so long without appreciating the outsized value of peatlands is both scientific and cultural. No one thought that peat could accumulate in the tropics because it was assumed that decomposition occurred too quickly in hot, humid weather to allow peat to pile up. It was Dutch and British colonizers who recognized that the ground they were standing on in Southeast Asia resembled the peat they harvested back home. This may be the reason why that enormous, 56,000-square-mile peatland in the Congo went undiscovered until 2017,[34] and how the mangrove peat of Botum Sakor National Park in Cambodia went unnoticed until 2012.

There is still so much to learn, as scientist William Shotyk told me one afternoon while I visited him at his office at the University of Alberta. Shotyk has spent his entire career studying peatlands in places such as the Jura Mountains of Switzerland, the Lindow bog near Manchester, England, and in bogs in the oil sands region of northern Alberta. The list of fens and bogs he has visited was so long that I stopped writing them down as he rhymed them off. "So many bogs, so little time," he said, urging me to visit the Isle of Arran in Scotland where you can hike up the rugged mountain landscape of Goatfell and visit the stone circles of Machrie Moor. And yet Shotyk still does not understand how these copper-deficient peaty soils produce plants such as cranberries and Labrador tea that are loaded with copper.

Nor did anyone know that pitcher plants trap and eat juvenile salamanders until Patrick Maldowan, an ecologist at the University of Toronto, saw one doing just that in the peatlands of Algonquin Provincial Park in Canada in August 2018.

We know that the reindeer lichen that I admired so much on my kayak trip down the Mackenzie River can be a dominant vegetation cover within peatlands such as those in the Ramparts Wetlands and the vast Hudson Bay Lowlands. But only recently has scientist Lorna Harris learned that reindeer lichens are in some ways more "super" than sphagnum, the "supermoss" of peatland ecosystems. In the first-ever study examining the effects of lichen on peat production and decomposition processes, Harris and her colleagues found that those thick lichen mats that caribou depend on so much in winter alter the vegetation composition of peatlands, reducing sphagnum cover and inhibiting the growth of small shrubs.[35]

These and other stories underscore the fact that we tend to focus on what is high on the food chain rather than what is low on the ground—or beneath it. The world is losing peat just as fast as the Arctic is losing sea ice. We're already beginning to see what a world without Arctic ice will look like. With sea ice melting, glaciers receding, sea levels rising, and Arctic storms picking up steam and putting a wrench into the jet stream that manufactures and moves weather from west to east, none of it is pretty.

The difference between Arctic ice and peat is that there are proven ways of growing or accumulating peat and promising research that may lead to better ways of restoring badly degraded peatlands and to persuasive arguments for keeping virgin peatlands in pristine condition.

Chapter 1

The Great Dismal Swamp

In the dark fens of the Dismal Swamp
The hunted Negro lay;
He saw the fire of the midnight camp,
And heard at times a horse's tramp
And a bloodhound's distant bay.
— "The Slave in the Dismal Swamp," Henry Wadsworth Longfellow, 1842

Along coastal Virginia where the British colonists began settling permanently in the spring of 1607, much of the land was tidal marsh, swamp, bogs, and fens filled with carnivorous plants and giant Atlantic white cedar and longleaf pine trees. It looked like paradise to those who had sailed for four and a half months across the cold, foggy ocean. "Faire meddowes and good tall trees," observed colonist George Percy. "I was almost ravished at the first sight thereof."

Almost immediately, the colonists began cutting down the white cedars that dominated the overstory and draining the bogs to produce bog iron. They relied on the trees to build cabins and to produce "tryals of Pitch and Tarre" to send back home. Pitch was used to seal boat hulls. Tar was used to grease the wagon axles. Britain badly needed both because most of its pines had been felled.

The terrain was presumably less wet and spongy than it typically would have been, due to an epic seven-year drought that started a year before they arrived. (Modern dendrochronology confirms this.)[2] This was known as the "starving time." Most of the crops planted by the Jamestown colonists in those first few years shriveled up by the end of summer, not necessarily for lack of farming skills, as some historians have suggested, but because of both the drought and also frost that came often during the abnormally cool period known as the Little Ice Age. The cooling began sometime around AD 1300 and intensified in the decades before and after the colonists arrived. River water and groundwater that was already brackish would have been more so as the drought persisted.

"There were never Englishmen left in a foreign Country in such misery as we were in this new discovered Virginia," Percy wrote in September 1607, just four months after he and 104 other people arrived with so much optimism. "We watched every three nights, lying on the bare cold ground, what weather soever came [and] warded all the next day, which brought our men to be most feeble wretches. Our food was but a small Can of Barley sod in water, to five men a day."

When a ship with supplies and another 100 settlers finally arrived in the winter of the new year, only 38 of the original colonists were still alive. Most of the others had died from a combination of malnutrition, salt poisoning, and typhoid. Typhoid alone killed more than 6,000 settlers between 1607 and 1624.[3]

On a rainy-day drive along the Pocahontas Trail from Williamsburg toward the Jamestown settlement, I saw very little of that spongy peatland ecosystem still intact. What I saw instead were badly degraded, peopled landscapes such as Busch Gardens, a sprawling outdoor amusement mall with theme parks and parking lots inexplicably named after England, Italy, and other European countries; joy rides called the Loch Ness Monster, Alpengeist, and Verboten; and a Rhine River cruise boat that plies the waters of a small lake. Farther down the Trail, and along other routes, I passed shopping malls, golf courses, residential neighborhoods of one-family houses and mobile homes, William and Mary University, and the Powhatan Resort manor, which dates back to 1735.

The resort and also a creek that cuts a verdant swath through the best-preserved forested peatland of James City County were named after the home village of a powerful North American Indian chief, known to the English as Chief Powhatan, who stood as head of an Algonquian-speaking empire. The territory, called Tsenacommacah, extended from the Potomac River in northern Virginia to the Great Dismal Swamp along the North Carolina border where the Nansemond, one of 30 or so tribes of Tsenacommacah, lived along the river that passes through present-day Suffolk.

Chief Powhatan, known to his own people as Wahunsenacawh, was the father of the legendary Pocahontas. All that is left of those people is a tiny reservation an hour's drive north of Jamestown.

I never fully appreciated the story of Pocahontas, who may or may not have saved the life of Captain John Smith, one of the leaders of the Jamestown colony. In my youth, Pocahontas was presented to us as a saccharine caricature in a Disney film. The dubious "facts" of her life were offered up in American history classes along with topics such as the establishment of the first permanent colony at Jamestown, the birth of American slavery a short time later, the networks of abolitionists such as Harriet Tubman and conductors of the Underground Railroad, and the Civil War that followed. Pocahontas was more myth than reality.

I did not see what connected Pocahontas and her people, enslaved persons, the Underground Railroad, and the Civil War until I drove from Jamestown to the Great Dismal Swamp, a vast peatland that stretches out along the border of Virginia and North Carolina. Lake Drummond, one of only two natural lakes in Virginia, sits in the middle of it.

This connection, which knits together so much history of the eastern United States, became clear as I explored the Great Dismal Swamp, both by myself and in the company of scientists who are trying to restore or sustain what's left of its peat. Seeing the fragments of land, and talking with the people who know it well, reveals the sharply contrasting points of view of the indigenous people who lived and hunted in these bogs, fens, and swamps, the fugitive slaves and social outcasts who took refuge in them, and the colonists such as Virginia-born William Byrd, who grew up in Britain and viewed the Great Dismal, when he returned to take over his father's estate, as a "horrible desart," a "blot on his Majesty's Kingdom," and a "miserable morass where nothing can inhabit."

"Never was rum, that cordial of life," he wrote, "found more necessary than it was in this dirty place."[4]

Eric W. Sanderson, the landscape ecologist who puzzled out what the island of Manhattan might have looked like before human settlement, once observed that people fit the environment to what best suits them rather than adapting, as animals do, by taking advantage of features of the environment which best suit them. Indigenous people like the Algonquin of the Powhatan confederacy were the exception. They made do with what peatlands like the Great Dismal Swamp offered rather than subdue the land, as the settlers tried to do in the centuries that followed.

Pocahontas was born in 1596, long before William Byrd set foot in the country. "Pocahontas" was a nickname meaning "playful one" or "ill-behaved child." Her birth name was Amonute. More formally, she went by the name of Matoaka, which means "flowers between two streams." She and her people lived comfortably in longhouse villages built on sandy escarpments and grassy *pocosins*, the Algonquian word for "swamp on a hill." Invariably, these settlements were situated along rivers and streams like the Chickahominy and Nansemond, tributaries of the James River. Peaty forests and wetlands like the Great Dismal, as well as "smaller dismals" like the Pitch and Tar Swamp at Jamestown, were places where they and other tribes went to hunt deer, bear, beaver, wild turkey, and ducks, and to pick nuts and fruits such as persimmons, pawpaws, and blackberries. Log fern (*Dryopteris Celsa*), fungi, and other forested peatland plants were harvested for their medicinal properties.

Young women like Pocahontas were taught to identify these and other plants. Pawpaws and log fern, they learned, were two sisters who grew together, one—the pawpaw—above the other—the log fern. Both were found in wettish, well-drained soils bordering swamps in neighboring uplands.

Powhatans grew tobacco for pleasure and ceremonial purposes, and maize (corn), squash, and beans for food. As modern gardeners know, these are the Three Sisters in agriculture. Corn offers support for beans to climb and grow. The beans pull nitrogen from the air to the soil to benefit corn and squash. The large prickly squash leaves ward off predators and shade the soil, keeping it moist and cool. It's important to know, because most peatland soils do not offer much in the way of nutrients for plants to take up.

The Powhatan tribes were so well adapted to what these peatlands had to offer that the first wave of British colonists often called on them to provide food during those "starving times" in exchange for tools such as hatchets and more fanciful items like the glass produced by Polish artisans, people not typically associated with that first wave of Jamestown settlers. The Poles and some Dutch people who were also brought along as skilled workers would produce not just glass, but pitch, tar, bog iron, and other materials. The Dutch, no doubt, were skilled at picking out lands that could be drained for the planting of crops.

Through trial and error, the colonists discovered what the New World had to offer. "If it not be ripe, [persimmon] will drawe a man's mouth awrie with much torment," noted Captain John Smith, who was charged with mutiny on that first sailing before party leaders learned upon their arrival that the Virginia Company had made him a member of the governing council. "But when it is ripe, it is delicious as an Apricocke."

More often than not, the early colonists puzzled over insect-eating pitcher plants, thorny "devil's walking sticks," venomous snakes, and vicious biting flies. Flies and snakes, red wolves and panthers may have deterred them from venturing too deep into the swamplands to harvest root vegetables and muscadine grapes or shoot the very many rabbits, deer, and other animals that dwelled there.

The friendly relations turned increasingly violent when the Powhatans could not, or would not, keep up with the colonists' incessant demands for food. The drought and cool weather were affecting yields from their crops as well. One Powhatan leader said as much when he called on Captain Smith "to pray to my [Smith's] God for raine, for their [the Powhatans'] Gods would not send any."[5] The colonial leader didn't believe him. He suspected that the Powhatans were simply hoarding food to pressure the colonists into leaving.

Deadly skirmishes followed as the situation escalated into an all-out war until 1613, when Captain Samuel Argall, the sailor who brought supplies to Jamestown, kidnapped Pocahontas "for the ransoming of so many Englishmen as were prisoners with Powhatan: as also to get . . . armes and tools . . . [and] some quantities of Corne, for the colonies relief."

Chief Powhatan offered to negotiate a price for the return of his daughter, but without success. In captivity, Pocahontas converted to Christianity and married colonist John Rolfe, the first to plant tobacco as a commercial crop in America. They then sailed off to England. She died there in 1617, by then known as Lady Rebecca. Rolfe, by some accounts, loved his wife, but he also allowed that he married her to induce Chief Powhatan and his tribes to be more cooperative.

Rolfe returned to Jamestown to manage his tobacco plantation. His fortunes, and those of the other struggling colonists, took a significant turn in August 1619 when a Dutch ship arrived with cargo in tow. Rolfe was initially disappointed. "It brought not anything but twenty and odd Negroes." Merchant Abraham Piersey and Governor George Yeardley, however, saw in these Africans a source of free labor for the planting of commercial crops such as tobacco and hemp, both of which were in high demand in Europe. They bought the Africans in exchange for "victuals"—food supplies that the Dutch captain badly needed.

These twenty enslaved people were among the first of 12.5 million Africans who were kidnapped, chained, and shipped off to America over the next two and half centuries. This forced labor allowed the colonists to drain swamps, bogs, and fens and greatly expand the planting of tobacco, hemp, rice, and eventually cotton, which, by 1803 had become America's leading export.

There was a limit, however, to what could be done locally. Because tobacco takes up more nitrogen, phosphorus, and potassium than almost any other plant, the soils around the Jamestown region became increasingly depleted of nutrients. Planting the "Three Sisters" might have solved the problem, but second- and third-generation colonists were more interested in making money from tobacco than growing food beyond what was needed to get them through a winter. Bogs, fens, and swamps that sat on sand, soil, and peat suitable for planting new crops were systematically drained until most of them disappeared.

It was William Byrd who suggested that the future of tobacco, hemp, rice, and corn lay in the Great Dismal Swamp. The plantation owner came to that conclusion after venturing into the Great Dismal Swamp in 1728 as one of the leaders of a survey to establish a disputed boundary line between Virginia and North Carolina, a state he mockingly dismissed as "Lubberland" and whose people, he said, "devour so much Swine's flesh that it fills them full of gross Humours."

Most everything we know about Byrd, including a painting in which he is dressed as an aristocratic dandy in blue velvet, suggests that he didn't have the ruggedness to survive an expedition as arduous as that one would have been. I met a few resentful Carolinians who suggested he was only there at the beginning and end of the journey. The truth is that he and other commissioners went around the perimeters of the swamp, leaving the surveyors to do most of the bushwhacking.

"A well-bred gentleman and polite companion" is how Byrd liked to describe himself. And that's how historians portrayed him well into the twentieth century. Byrd, however, was anything but a gentleman, to judge from entries in secret diaries that were overlooked by many historians and not published until 1942.[6] There was no end to the number of times he boasted about "rogering" his wife (on a billiard table, no less), often with a "flourish" following what was frequently a "little quarrel." He bedded maids and enslaved women, sometimes against their will. He was sorrier for sometimes forgetting to say his prayers at night than for using "those women as if they belonged to me when they really do not."

His idea of a joke would have brought the Pilgrims at Plymouth Rock to their knees in prayer. Letters to an old school friend were addressed to a "Mr. Cocke" and signed by "Mary F-x." He also beat and whipped the people he enslaved anytime it pleased him, recording the myriad cruelties with seeming satisfaction in his diary.

Sadist and misogynist though he was, Byrd was also a serious naturalist who consulted with scientists about plants and animals and the swampy peatlands that clearly fascinated him. His descriptions of the Great Dismal Swamp were often spot-on.

"A boggy place" in which "white cedar," is commonly mistaken for juniper, " he wrote, explaining why some called part of the Great Dismal "Juniper Swamp."

"The reeds which grew about 12 feet high, were so thick, & so interlaced with Bamboe-Briars, that our Pioneers were forc't to open a Passage. The Ground, if I may properly call it so, was so spungy, that the Prints of our Feet were instantly fill'd with Water."

Byrd ate the meat of bear and venison mixed in stew and seemed to enjoy it. He saw panthers and heard red wolves howling "with the yell of a whole family of wolves, in which we could distinguish the treble, tenor, and bass, very clearly."

Byrd, however, also crafted an image of bleakness in the Great Dismal that evoked the most macabre of medieval superstitions about swamps and bogs being sources of poisonous vapors rising from rotting vegetation. A "miasma," as it was called in those days.

Birds, he claimed, didn't fly over the Dismal Swamp "for fear of the noisome exhalations that rise from this vast body of dirt and nastiness. These noxious vapours infect the air round about, giving agues and other distempers to the neighbouring inhabitants."[7]

The best and profitable thing to do, he advised, was to drain it. "By draining the Dismal," he wrote, "it will make all the adjacent country much more wholesome, and consequently, preserve the lives of many of the king's subjects: this will happen by correcting and purifying the air. . . . After the Dismal comes to be drained, it will be the fittest soil in the world for producing hemp, the propagating of which, is with so much reason, desired and encouraged in his majesty's plantation."

Byrd drafted a proposal to members of the Council of Virginia calling for the formation of a company to drain the swamp. He envisioned using enslaved laborers to build roads and drainage ditches in order to set the stage for agriculture. Huckster that he was, he expected a return on investment of 900 percent within ten years.

Virginians searching for new places to expand tobacco farming along with other crops may well have acted on his advice had they not had bigger challenges to contend with at the time. The threat of a Powhatan attack was still a constant worry, given the fact that there were three periods of war in Virginia between 1609 and 1642 as well as several nasty skirmishes in the decades that followed.

The growing number of enslaved people, who weren't nearly as compliant as they was expected to be, were also a growing concern. By

1740, there were 60,000 enslaved people in the colony. Runaway slaves from Virginia tobacco plantations and South Carolina rice fields routinely took refuge in swamps and pocosins, emerging at times to steal food from the plantations they had once worked on. At one point, militiamen were encouraged to take their guns to Church on Sundays in the event that fugitives might come out of those swamps and attack parishioners.

The worst of these fears was realized in 1739 when an educated slave by the name of Jemmy (Cato in other references) recruited as many as 80 enslaved people in a South Carolinian rebellion that came to be known as "The Stono," after a river in the region. Twenty-five colonists and as many as 50 slaves were killed.

Byrd's idea of draining the swamp, however, was not forgotten. In 1753, the Virginia Council granted four men a large portion of the Great Dismal after concluding that it was "at present altogether useless." One of those land speculators was Robert Tucker, a mercantile trader, slave owner, and former and future mayor of Norfolk County. Over the course of the next decade, he amassed more land and an even larger group of wealthy investors that included George Washington, the future president of the United States.

In 1763, the Dismal Swamp Land Company, sometimes referred to as the Adventurers for Draining the Dismal Swamp, was formed to do what Byrd had recommended 35 years earlier. Washington visited the Great Dismal Swamp seven times, charging expenses to the company in each case. Suitably impressed by what he saw and shot—Washington loved to hunt ducks—he put together a labor force that included 54 enslaved people rented out by the investors. Nine women, a boy, and a girl were among them. The ditch they dug bears his name to this day.

In an effort to add efficiency to the ditching, and to ward off a rival company that was buying up land, Washington at one point contemplated bringing in 300 Dutch or German laborers "who were more acquainted with draining, and other branches of agriculture."[8] The Dutch had been refining the art of draining wetlands since 1050, when counts and bishops began offering people freedom and land ownership opportunities so long as they made the wetlands "habitable." Their skill in draining swamps, bogs, and fens was in great demand in England,

and their reputation had extended to America as well. But instead of doling out more money to bring in the Dutchmen, investors in the company blamed each other for offering up enslaved people who, according to Washington, were "not up to the task of subjugating the Swamp."

The draining and digging of ditches continued with those enslaved men and women when Virginia Governor Patrick Henry ("Give me liberty or give me death") came up with the idea of constructing an enormously long canal that would connect Albemarle Sound in North Carolina with the Chesapeake Bay. Washington was president at the time and reluctant to back the project until Thomas Jefferson got him to change his mind.

When the 22-mile-long canal was finally completed in 1805, it wasn't deep enough to float the bigger boats that were needed to haul out roofing shingles and timber—products that had become a much bigger source of profit than rice. So, for the next seven years, more enslaved people were brought in to dig the Feeder Ditch (the Jericho Canal) in order for water levels to be raised.

There was no end to the ditching and draining. Lotteries were staged to raise funds for improving the canals in 1819 and 1820. In 1826, the US Congress bought shares in the Dismal Swamp Company after recognizing the strategic importance of the country's waterways. As this new era of canal-building unfolded all across the United States, President Andrew Jackson visited the Great Dismal in 1829, setting the stage for Congress to buy even more shares. The investments allowed for troublesome squared-timber locks to be replaced by granite hauled in from New Jersey.

Working in the buggy, venomous swamp was unquestionably a nightmare. Moses Grandy, an enslaved person who was consigned to work as a boatman, described ditch-digging in a book he wrote after gaining his freedom:

> The ground is often very boggy; the negroes are up to the middle or much deeper in mud and water, cutting away roots and baling out mud; if they can keep their heads above water they work on. . . . No bedding whatsoever ever is allowed them; it is only by work done

> over this task that any get a blanket. They are paid nothing, except for this overwork. Their masters come once a month to receive the money for their labor. . . .[9]

The swamp, however, was sometimes a shortcut to freedom and an end to the violence that was often delivered on a plantation. In an ad, typical of many taken out in Virginia and Carolina newspapers, a plantation owner offered a reward for the return of a runaway and described him this way: "My man George; has holes in his ears; is marked on the back with a whip; has been shot in the legs; has a scar in the forehead."[10]

George may well have joined a community of fugitives that had settled in the Great Dismal, assuming that enslavers would have a tough time getting in by horse or foot to retrieve him. Fugitives like these were known as "maroons," from the Spanish word *cimarron*, meaning wild or untamed. The word was also used to describe fugitives in Jamaica, Haiti, the Everglades, and elsewhere in many southern states.

The Great Dismal's fugitives stayed alive by adapting, as the indigenous people did, to what the environment had to offer. They grew corn, rice, and wheat. When the crops were harvested, they picked nuts, fruit, and berries, dug up root vegetables, fished for bass, crappies, and pickerel that abounded in Lake Drummond. They hunted wild hogs and deer. They also traded with social outcasts—poor men and women who did not fit into White society—and with members of the Tuscarora and the Nansemond—Algonquian-speaking people who lived in and along the fringes of the swamp.

Thanks to the merchant records recently unearthed by historian Marcus P. Nevius, we know that one fugitive lived in the Swamp for 13 years before emerging sometime in the early 1770s. Those records show how this maroon "raise[d] Rice & and other grain" to sustain himself and by building furniture that he sold. The merchants make no mention of the buyers, but they are evidently impressed that a fugitive could somehow survive tolerably well in a refuge for "many Bears, Tigers, Racoons," and a "great lake."[11]

As Nevius points out, scholars like him have faced a "deafening archival silence" in attempting to document human history in the Great

Dismal Swamp as anything more than a place where enslaved people dug ditches and where maroons stopped for a while to hide and rest before moving on.

Archeologist Daniel Sayers has unearthed artifacts that reveal a much more complex reality; diasporic communities formed in the Swamp as thousands of formerly enslaved people migrated in and out. Subsistence patterns set in, and exchange systems were developed, not just within the maroon community but with the outside world as well.[12]

Some of this we already knew thanks to those merchant records, William Byrd, and many others who encountered fugitives living in the Great Dismal over the course of two and a half centuries. Byrd, for example, reported seeing "a family of mullatoes" in the Great Dismal when he was surveying the area. "It's certain many slaves shelter themselves in this obscure part of the world," he noted.

Maroons were there in the early 1780s when John Ferdinand Smyth passed through the region, horrified at times by "insects, reptiles, and serpents of the most poisonous, deadly, and fatal nature" that he saw along the way. "Run-away Negroes have resided in these (swamps) for twelve, twenty, or thirty years and upwards. . . . These horrible swamps are perfectly safe, and with the greatest facility elude the most diligent of pursuers."[13] Smyth's description of "wild beasts howling hideously," in "dreadful" overcast conditions where rain seemed to be a constant threat, further underscored the subliminal intimidating nature of the Great Dismal.

Irish poet Thomas Moore ("*Thee, thee, and only thee*") added to the mystique when he visited the region in 1803, penning a ghostly poem about a legendary native American bride-to-be who disappears just before she is about to be married. Her lover goes mad, believing she has ventured into the Great Dismal before losing her way.

"They made her a grave, too cold and damp
For a soul so warm and true;
And she's gone to the Lake of the Dismal Swamp.
Where, all night long, by a fire-fly lamp,
She paddles her white canoe.
"And her fire-fly lamp I soon shall see,

And her paddle I soon shall hear;
Long and loving our life shall be,
And I'll hide the maid in a cypress tree,
When the footstep of death is near."

American interest in the Great Dismal grew from there. Duels were fought there. Lovers came for the romance conjured up in Moore's poem while hoping to get a glimpse of that ghostly maiden floating across Lake Drummond in a white canoe at night. Henry Wadsworth Longfellow wrote a poem about it. Edgar Allan Poe's "The Sphinx" was inspired by the miasmic nature of the place. Poe may even have finished writing "The Raven" while staying at the Lake Drummond Hotel.

Capitalizing on this interest, the editors of *Harper's New Monthly Magazine* sent Virginia author-illustrator David Hunter Strother to the Great Dismal in 1853. It was in some ways an odd choice. Strother was best known for drawing racial stereotypes of Black cooks and tidelands fishermen, not for his social commentary. In truth, he was not so much interested in runaways as he was in getting a glimpse of the lake where that ghostly bride was seen on misty evenings

Strother got a frightful start when he heard something moving toward him in one of the more remote parts of the Swamp. "I paused, held my breath, and sunk quietly down among the reeds," he later wrote. "About 30 paces from me I saw a gigantic negro, with a tattered blanket wrapped around his shoulders, and a gun in his hand. His head was bare, and he had little other clothing than a pair of ragged breeches and boots. . . . The expression of the face was of mingled fear and ferocity, and every movement betrayed a life of habitual caution and watchfulness. . . . Fortunately, he did not discover me, but presently turned and disappeared."[14]

Like most White people, Strother couldn't shake off the notion that vast peatland swamps like this were uninhabitable places of mystery where ghosts and murderous criminals carrying jack o' lanterns lurked and where maroons like the giant slave he saw were ready to pounce on anyone who tried to find them.

Frederick Douglass, himself once a fugitive slave, explained why when he compared the Great Dismal to Canada, the final destination for many maroons. "It is the dreariest and the most repulsive of American possessions. It is the favorite resort of wild animals and reptiles, the paradise of serpents and poisonous vegetation. No human being, one would think, would voluntarily live there; and yet . . . it has been the chosen asylum of hundreds of our race . . . the healthy haven of negro slaves where the wicked cease from troubling, and the weary are at rest."[15]

A maroon named Charlie said as much when American journalist James Redpath interviewed him in 1854. "Da calls it Dismal Swamp," he allowed. "And guess good name for it. Tis all dreary like. Dar never was any heaven's sunshine in some parts orn't."[16] But that was not his point. "Best water in Juniper Swamp ever tasted by man," he said. "Dreadful healthy place to live, up in de high land in de cane-brake."[17]

Although some of the maroons were caught, many women, children, and men moved about with such stealth that even their nighttime raids on the nearby farms of slave owners went unnoticed until it was too late to track them down. More often than not, these raids were timed with the aid of critical information provided by enslaved people—friends and family members, most likely—who kept in touch with the runaways by various means, including leaving useful tools such as scythes in the fields for them to pick up.[18]

"Dar is families growed in dat dar Dismal Swamp dat never seed a white man, and would be skeered to def to see one," said Charlie. "Some runaways went dere wid dar wives, an' dar childres are raise dar."[19]

Frederick Law Olmsted, the journalist who later designed New York's Central Park after thoroughly draining it, visited the Dismal Swamp in January 1853 as the growing number of fugitive slaves was becoming a subject of intense national interest—anathema to plantation owners, and a siren call to slaves who were desperate to be free, and abolitionists like Frederick Douglass who wanted to free them. Untold numbers of rebellious maroons by then had escaped not only to the snake-filled peatlands of the Great Dismal, but to swamplands along the Savannah River in South Carolina and the Everglades in Florida, where they joined forces with Seminole Indians in a war with slaveholders that lasted from 1835 to 1842.

Olmsted arrived at a time when all that was left to remind people that this land was once Tsenacommacah was a war steamer named after Chief Powhatan. Olmsted found Norfolk to be "a dirty, ill-arranged town nearly divided by morass (boggy ground)." His description of the Great Dismal was one of profound degradation. "Nearly all the valuable trees have now been cut off from the swamp," he wrote. "The whole ground has been frequently gone over, the best timber selected and removed at each time, of course leaving the remainder standing thinly, so that the wind has more effect on it, and much of it, yielding of the soft, is uprooted and broken off."[20] Fire, which was once rare in the Great Dismal, was by then a frequent visitor, according to Olmsted, because draining so much land left peat exposed to the drying effects of the sun."[21]

And yet it was a still a busy place. Olmstead mentions a proprietor who employed more than "one hundred hands" transporting shingles out. Moses Grandy recalled seeing as many as 700 enslaved people digging ditches.[22] Olmsted suspected that the days of making money from timber were coming to an end because "there is little or no large wood growing to supply the place of that taken off, except in the drier parts where pines come up in 'old fields.'"[23] He suggested that it would then become "dead property," unlikely to become good farmland unless another extensive scheme of draining took place to reveal new soil.

Olmsted also saw that the degradation brought on by clear-cuts and drainage would increase the vulnerability of a maroon who had settled into the Dismal as a free man who, "having the liberty of the swamp, hunts coons, fishes, eats, drinks, smokes, sleeps and works according to his will."[24] Many fugitive slaves must have sensed this as well. The world had closed in on them when the Fugitive Slave Act was passed in 1850. The law compelled even free states to return enslaved people to their enslavers. The slave hunters, Olmsted noted, used "Bloodhounds, fox hounds, bull dogs and curs' to track down the runaways."[25]

When Olmsted asked his guide if he knew of maroons being shot after being tracked down and cornered, the man acknowledged that this was so. "Oh yes," was his reply. "But some on 'em would rather be shot than be took, sir."[26]

Far from being the inhospitable wasteland described by White colonists, there was bounty and beauty in the Great Dismal, even if it took grit to survive. It is a testament to the vast wilderness itself, as well, that so much could survive the onslaught from profiteers bent on the Swamp's destruction. And, increasingly, it was difficult to profit from draining the land.

Some maroons eventually found their way to Canada with the help of a coalition of abolitionists such as Harriet Tubman and Canadian doctor Alexander Milton Ross, who used his birdwatching and butterfly-catching hobbies as a cover to shepherd fugitive slaves from the South to points north. Ross would knock on the doors of plantation owners and ask if he could look around for birds and butterflies, making no mention of the fact that he was really there to interview and inform slaves working in the fields of possible avenues to freedom. Ross would often give them money, a pistol, and directions for how to get to safety.

Maroons who didn't find freedom by land did so by hitching rides on ships that passed along coastal regions of the swamp. The captains of those boats often came ashore looking for fish, meat, timber, and fresh bog water, all of which the maroons used for barter. In exchange for these items and/or money the maroons earned by working in an underground economy, they purchased passage to safe havens farther north in places like Seneca, Virginia, where abolitionists as well as black watermen who worked on those ships knew which captains were likely to take the maroons aboard without turning them in.

The final chapter of the Great Dismal Swamp Land Company came six years after the Civil War ended, when partners leased their entire holdings to Baird, Roper and Company, who soon found themselves in competition with several other investors who had bought up land specifically to drain the wetlands so that they could get whatever timber remained. Some of these shares were sold by Congress, which unloaded its investment in the Great Dismal Canal in 1878.

Author Charles Frederick Stansbury predicted the demise of the Great Dismal in an account of his journey through the Swamp in the early 1920s. "If left alone by legislation which now threatens to drain it, for the purpose of reclaiming its broad acres for the use of husbandry, the Dismal Swamp may for a century remain a delight to visitors with its beauties and its mystery.

"But even so, the woodsman's ax is slowly eating away to the shores of Lake Drummond. . . . The swamp will be shorn of its tree growth, and where now beauty and interest abide at every step, only the blackened specters of dead trees will stand sentinel over its denuded and burned-over surfaces.

"The time to see the swamp in its primeval loveliness is now and in the near future, for like all things beautiful, it must pass away before the encroaching commercialism of the times."[27]

Time proved him to be wrong. The logging continued until most of the mature trees were gone. But Union Camp, the company that eventually acquired the shares in the Great Dismal Swamp Company, transferred 50,000 acres to the Nature Conservancy in 1973. The Nature Conservancy turned it over to the US Department of Interior the next year for the creation of the Great Dismal Swamp National Wildlife Refuge.

The peatlands of the Great Dismal Swamp began forming around 9,000 years ago in coastal depressions between relic dunes left behind during the Last Ice Age, when the Atlantic shoreline receded. Fibrous peat from the slow decay of water-loving plants accumulated at rate of about an inch each year on a clay foundation that limited downward drainage of rain and groundwater flowing in from the Suffolk Scarp to the west. The peat is more than twelve feet deep in some places, less so in most others. Just 20 percent remains of the million or so acres of forested peatlands that existed before William Byrd and his fellow surveyors mapped the region.

On the morning I arrived in the Great Dismal Swamp, Fred Wurster was pointing out places on a map where the maroons once lived. Wurster is a hydrologist working for the US Fish and Wildlife Service. With him was Eric Soderholm of the Nature Conservancy, which helps fund ecological restoration efforts in the Great Dismal, the Pocosin Lakes National Wildlife Refuge to the south, and in other North American peatlands.

Wurster and Soderholm weren't so much interested in providing me with the story of the maroons as they were in describing what is being done to rectify the ecological degradation that has been wrought by two centuries of deforestation and the draining and diversion of so much water.

"There are more than 150 miles of roads, canals, and ditches in the refuge, many of them going back to the days of George Washington," Wurster told me as we headed into the heart of the Swamp. "These and other disturbances have disrupted the natural flow of water. Many of the wetlands dried out as result. Peat holds water that is needed to prevent wildfire and flooding. It also stores carbon that would otherwise contribute to greenhouse gas emissions."

The biggest of the drainage channels is the Dismal Swamp Canal. When it was opened to traffic in 1805, it effectively stopped the eastward flow of water in the Swamp. Land east of the canal gradually dried up. Lands west got a lot wetter. Fire on one side and flooding on the other emerged as constant problems. Other smaller ditches created similar scenarios, but in different ways.

As we drove along dirt roads and grassy pathways, I stared out the window of the truck at patchy groves of cypress, gum, and swamp bay, a tree that can grow up to fifty feet high and twelve feet in circumference. Several of the Swamp's 200 species of birds eat the bitter fruit of this tree. Caterpillars of the Palamedes swallowtail butterfly, which has a wing span of almost half a foot when it morphs into a flyer, consumes its leaves.

Most everything is big here, I discovered. There are heavy-bodied canebrake rattlers that are up to five feet long and black rat snakes that are longer than the tallest NBA basketball player is tall. Horseflies are twice the size of my thumbnail. The bite of the slightly smaller yellow fly is so excruciatingly painful that the Refuge posts a video on Facebook warning visitors that bug spray is often insufficient to prevent bites, and that a bug hat and full clothing cover are advisable.

There are also ticks. Some of the first cases of Lyme disease in the United States were reported by people who lived near the swamp in the early 1970s. Many came down with mysterious symptoms that included fatigue, confusion, and problems with hand–eye coordination. Local doctor Robert Bransfield, who went on to become a leading expert on Lyme disease, had a hard time diagnosing the symptoms when they were first brought to his attention. It was, he recalls, the "invisible disease."[28] Some doctors attributed the complaints of fatigue and confusion to mental health issues. Insurance companies refused to compensate victims for health care costs.

No wonder, then, that as recently as 1944, a local scholar described the Great Dismal as a "place of doom and mystery where the sun never reaches the ground, a place where the deadly cottoned-mouthed moccasin and his brother the rattler vie with Old Malaria and his two cousins, Lingering and Slow Death, in keeping Man out of one of Nature's last strongholds."[29]

The farther we drove into the Swamp, the more I realized how dangerous and easy it would have been to get lost before there were roads and canals. This is what happened to William Drummond, the colonial governor of North Carolina from 1663 to 1667, when he ventured in with a hunting party. As he and his companions wandered through the tangle of vines and brambly thickets, they encountered what Drummond described as a "Green Sea" of river cane, the only native bamboo that grows in the United States. They also found the inland lake where Strother was to see that giant slave more than a century and a half later. Drummond was fortunate, as he was the only one to emerge alive. (He didn't get to tell his thrilling story for long, however. He was hanged for treason in 1676 after leading a failed rebellion against the governor who succeeded him.)

Iconic photos of tupelo and bald cypress trees rising up from the shallow, tea-colored water along the shores of Lake Drummond are typically used to describe what the Great Dismal Swamp looks like. Clumps of Spanish moss draped along their branches give them the funereal look that Strother described so gothically. It is a striking scene that would have been even more haunting in the days when panthers climbed those trees and when snakes fell out of them as people and trains carrying timber passed by. On one occasion, a train engineer climbed into his cab to find a panther sitting on his seat.[30]

Like the red wolves that William Byrd heard howling in 1728, the panthers are long gone from the Great Dismal. Red maple dominates—not tupelo, longleaf and loblolly pine, Atlantic white cedar, or sweetgum. Most of the giant cypress and white cedars that Byrd, Strother, Olmsted, and others described are now mostly rotting stumps.

Lake Drummond: For more than a century, entrepreneurs like George Washington attempted to drain the Great Dismal Swamp with enslaved people. (Photo: US Forest Service.)

The shortage of mature trees represents a serious challenge, not just for butterflies like the rare Hessel's hairstreak caterpillar that feeds on the leaves of white cedar, but for efforts to reintroduce the critically endangered red-cockaded woodpecker. The nine-inch-long birds build their nests in cavities they peck out near the top trunks of 70- to 100-year-old longleaf and loblolly pines. Because there are so few of these mature pines left in the forested peatlands, refuge biologist Jennifer Wright and others carve nests into the best ones for chicks that have been plucked from other swamps in the Southeast. "It takes two to five years for the adults to excavate a cavity big enough to hold a nest," said Wright when we met up with her in the high ground of the pocosins area. "These birds don't come to the ground to feed, and they are also not great flyers. That's why they need help getting re-established."

We were standing in the middle of a trail that had just been cleared by a Swamp tracker. The weight of the machine had squeezed so much moisture from the spongy peat that water was still percolating to the surface. I had an uncomfortable hunch that if we had stood in that place for an hour longer, we would have been up to our thighs in mud and water, unable to get out.

Wright confirmed something I had been suspicious about when Wurster told me earlier that these birds peck holes around their nest to promote sapping. The sap, he said, is supposed to deter the giant black rat snakes from slithering up the trunks to the tree tops and eating the eggs and chicks. Coming from a boreal world where the hopelessly timid garter snake, which is usually no more than three feet long, is as intimidating as reptiles get, it was difficult coming to grips with an image like that, despite how nonvenomous and important to the ecosystem these snakes are.

When we drove on and stopped at another point along the way, Wurster had a pair of snippers in hand to provide us with safe passage through the jungle of tangled vines, spiky devil's walking stick, and prickly thickets of blackberry. I understood why he and Soderholm were wearing thick canvas pants on this warm day when I brushed up against one of the blackberry bushes. Grabbing onto rosehips in a boreal bog, as I have sometimes done to balance myself while getting out of a canoe, can be mildly painful. But here, my denim jeans were not enough to prevent blood from being scratched out of my thighs.

"See how dry it is," said Wurster as he pointed to the forest floor. "Before the canals and ditches were dug, the bottom trunks of those trees would have been under a foot or more of water. When water was drained out, the peat dried up and the ground subsided. That's why so many of the roots you see here are now exposed. When the wind blows hard many of these trees fall down."

The biggest of these knock-downs occurred in 2003 when Hurricane Isabel, one the most powerful storms to reach the Chesapeake Bay area since 1933, felled most of the 4,000 acres of the last pure stand of Atlantic white cedar in the Swamp. A sliver of the forest survived, and a massive restoration effort was put into play in 2006.

Wurster was a latecomer to this ever-expanding restoration effort when he arrived in 2010. He had previously worked out of the Pacific regional office in Portland, Oregon. While he was there, he offered his expertise to refuge managers in Washington, California, Idaho, and

Nevada. He also evaluated wetland habitats in desert landscapes such as the Mojave, where another Fish and Wildlife refuge, Ash Meadows, is home to a remnant peatland swamp dating back at least 8,000 years.

Wurster did his graduate work with David Cooper, a Colorado State University scientist who has done more for the discovery and understanding of alpine fens than perhaps any scientist alive today. Many national parks, such as Yosemite, didn't know they had alpine fens until he showed them where they were and why they were important to everything that lived there and down below.

For Cooper and many of his colleagues and students like Wurster, the key to understanding and appreciating peatland ecosystems is hydrology—the ways in which precipitation and groundwater recharge peatland ecosystems before spring runoff and summer heat lower the water levels, sometimes draining them altogether.

Rain provides most of the water that goes into the Great Dismal; about 28.5 billion gallons fall annually. Groundwater that wells up through the peat makes a contribution, but not in the way it once did. Roads, tilled fields, and housing developments in the uplands along the Suffolk scarp that borders the west side of the swamp have cut off or altered the flow of this water. Factor in ditches and canals along with warmer temperatures that are accelerating evaporation, and you have a hydrological problem with enormous ecological implications for plants, birds, and animals, and for wildfire.

Since 1974, when the US Fish and Wildlife Service began managing the Swamp, various forms of water-retention structures have been built to prevent rainwater from draining out. Completely restoring this spongy peat to what it once was is not realistic, Wurster allowed. The plumbing in the swamp, for one, is too complicated.

It looks more like water pouring through a spaghetti cauldron than what one would typically see in uphill–downhill flow patterns. For another reason, it would require filling in the ditches and canals and removing all of the roads. Not that Wurster hasn't proposed doing some of that. It's a tricky proposition, though, because the roads were dug deep down into the peat and the clay below in ways that altered groundwater flow. Roads are also needed for human access and to help fight wildfires that have been burning bigger, hotter, and more often, further complicating the hydrology of the refuge.

"I know it sounds strange to have a wildfire problem in a swamp," said Wurster as we drove along a road where a wildfire burned for nearly four months in 2011. "I suspect we are one of the few places in the world where our firefighters wear hip waders. But we do have fires, and they are sometimes very difficult to put out."

The 2011 fire also left an indelible mark on the landscape. The scarred area is more swamp than pocosin. This is no longer habitat for bears, deer, racoons, or the southern bog lemming (first identified in the Great Dismal in 1898 and not rediscovered until 1980). Only now, nine years after the fire burned, are the Atlantic white cedar and pond pine seeds beginning to show signs of life. It's the heat of fire that induces pine and spruce cones to give up their tightly clad seeds. Without fire, these forests cannot regenerate naturally.

There is a long history of fire in the Swamp. Lightning storms cause many of them, more so now than in the past because of how much the peatland forest has dried out.

A year after the Great Dismal Canal was opened to traffic in 1805, a fire burned for nearly a month. Laborers tossed tens of thousands of roofing shingles into the water to save them from being destroyed. The shingles burned anyway when they floated back to the surface. Another fire in 1839 was thought to have been even bigger.

The fire situation, according to Wurster, worsened over time as the ditch system grew and more water drained out into the Great Dismal Canal, leaving the peat to dry and then smolder when ignited. These peat fires are particularly difficult to put out because they can smolder to a foot or more below the surface where no amount of water, save for a tropical storm, can douse them. They can also alter the landscape in perilous ways. In the 1880s, railroad workers working in the area thought they were walking on solid ground when they suddenly dropped out of sight into smoldering peat that had ignited. To ensure that trains passing through didn't suffer the same fate, locomotives were used to push carts full of stone ahead on the track so that the ground would have time to give way before causing more serious damage.

The Great Dismal Swamp has had peat fires that have persisted for months and, in some cases, years. A fire in 1923 smoldered for three years. Two fires in 1941 and 1942 clouded the coastline with so much smoke that American vessels patrolling the shoreline for German

U-boats during World War II had a difficult time navigating through the shoals. Lightning was usually the cause of these fires, but humans were also responsible. Sparks from a passing train in 1955 caused the Easter Sunday Fire that burned 150 square miles. A person burning debris caused another fire in 1967. The April Fools Fire burned when a prescribed fire got out of control in 1988.

It was two record-breaking fires—the one in 2011 and another that burned here and in the Pocosin Lakes National Wildlife Refuge three years earlier, that turned a lot of heads. More than 400 firefighters were brought in to put them out. Hospitals in the regions saw a huge spike in the number of people suffering from respiratory problems. The situation became so desperate that Wurster and his colleagues engineered a pumping system that sent water uphill from the ditches below in order to put moisture back into the tinder-dry peat. Even then, the fires continued to smolder.

Most everyone thought that the torrential rains that came with Hurricane Irene in late August of that year would have been enough to extinguish the peat fires. But the stormwater drained out through the ditches so quickly that parts of the swamp continued to smolder. "That was a real eye-opener," said Wurster. "We thought for sure the heavy rain would put out that fire. Fortunately, more rain came that fall."

Hurricane Sandy, which pummeled the swamp with record rainfall the following year, highlighted another side of the hydrological challenge that refuge managers like Wurster are dealing with. In this case, neither the Swamp's spongy peat nor the water-retention structures could soak up and hold back enough water to prevent the flooding of adjoining neighborhoods and farms. And that's when the shit hit the fan, so to speak. Someone had to take the blame, and the Fish and Wildlife Service was an easy target because the agency was not popular with some landowners who objected to the creation of the wildlife refuge in the first place. Philip McMullan Jr., a former farming consultant, echoed the sentiments of long-dead colonists when he said, "I can't see that there is any value of pocosins to the environment. It's an inert piece of nothing."

Flooding is a problem that millions of people, from New Jersey to the Florida Everglades, are increasingly facing at a time when sea

The Great Dismal Swamp is densely populated with black bears and many endangered species such as the southern bog lemming and the red-cockaded woodpecker. (Photo: US Fish and Wildlife Service.)

levels are rising while coastlines are subsiding in places where peat has been either removed or allowed to dry out. And as greenhouse gases that blow off from this peat add to the copious amounts of human-produced carbon that is already rising into the atmosphere, a rapidly warming climate is generating more tropical storm activity. The fire of 2011 sent as much carbon into the atmosphere as a million cars do each year.

To slow these greenhouse gas emissions and to mitigate the flooding that is likely to get worse, money from the Hurricane Sandy relief fund was allocated to build even more water-control structures in the Swamp and in other places along the low-lying coast. The hope is that rewetting the peat will reduce wildfire severity and increase the amount of carbon stored in it.

Wurster and Soderholm showed me one of the larger retention structures that has been constructed at South Martha Washington

Ditch, which runs parallel to the Dismal Swamp Canal. The first thing that comes to mind when people see tan-colored foam streaming several hundred feet downstream of the small waterfall below the weir is that this water had been polluted in some way. But the foam comes from the tannins in the tea-colored water.

Water-retention structures, even as big as this one, are easy to operate with wood or aluminum slats that can be added or removed in minutes. Thirty-four have been installed since Hurricane Sandy flooded the area, and there are plans to install about twelve more.

Wurster struck me as an especially calm person, given the fact that he is condemned if he releases too much water downstream into farmlands or holds too much of it back. "If you put too much water in wetlands, you end up with more methane emissions," he said. "Methane is roughly 20–25 times more potent than carbon dioxide. That will defeat the goal of slowing climate change."

Effective as these water-retention structures are, aluminum and cement don't exactly blend in with this woody landscape. So, with the help of the Nature Conservancy, there's an attempt to enlist the help of the Swamp's beaver population. This is being done by driving in wood posts across the length of the ditch and threading small branches between the posts to create a makeshift dam. The strategy? "If you (help) build it, they will come."

Soderholm showed me one such water retention structure they had recently set up on the northeast side of the swamp. "The restoration engineer we're working with suggested we give it a try because it worked in some places out West," Soderholm said. "We had some money left over from a tree-planting project, so we decided to do this as an experiment. There are beavers here, but we're still waiting for them to do their thing."

The idea of using beavers as engineers of ecological restoration is a relatively recent one, gaining in popularity. The US Fish and Wildlife Service even has a handbook on the subject. In England, where beavers were hunted out 200 years ago, a Beaver Advisory Committee has been set up to oversee the use of beavers to restore drained peatlands. A pair was recently dispatched to a badly degraded peatland site in Devon. In Scotland, beavers have been successfully reintroduced to

Knapdale, and unofficial populations are thriving in the River Tay and its tributaries.

Mother Nature's furry, flat-tailed plumber may someday come to the rescue in the Great Dismal. But there are many other problems that beavers can't solve. Farms bordering the Swamp use fertilizers and pesticides on corn, soybeans, cotton, and peanuts (which are boiled here, rather than roasted). And the preference for pork that goes back to William Byrd's day can be seen in hog operations that seem to be everywhere.

A bigger concern lies with lawmakers and the Supreme Court justices who have either rolled back or called into question legislation that protects peatlands such this. Potentially, the most damaging Supreme Court ruling came in 2006, when a Michigan man asserted his right to fill 22 acres of wetland to make way for a mall. Writing on behalf of the majority, Justice Antonin Scalia cast doubt on whether peatlands such as the Great Dismal (he did not mention it by name) can even be considered wetlands (and therefore entitled to legal protections) because they do not have surface connections to large bodies of water.

A few days after exploring the Great Dismal, I decided to treat myself to a short paddle through the Black River Preserve in North Carolina, home to one of the very few ancient bald-cypress--bottomland forests in the world. The wetlands there are as soggy as they are in the Great Dismal, and the forests are just as impenetrable. The water is as black as tea.

The southeastern bald cypress that grows here is the fifth-oldest tree species on earth. I found some of these gnarly, storm-battered specimens draped in Spanish moss in an area known as the Three Sisters. No one knew exactly how old these trees were until David W. Stahle and his colleagues went in and aged them with a nondestructive coring technique. Counting the tree rings, they found that one of them was at least 2,654 years old, making it the oldest living tree east of California. Only the giant sequoia, Sierra juniper, and Great Basin bristlecone pines of California, as well as the Alerce of Chile, are older.

The thing that sustains the bald cypress is the acidic, nutrient-poor swamp water. The trees grow slowly and achieve remarkable age because invasive species, natural competitors, and parasites have difficulty getting a foothold in these hostile conditions.

Bald cypress, I like to think, are a little like the tamaracks that also grow old, slowly and gracefully in the boreal forest, even when fire consumes huge chunks of their trunks. They do well in peat, just as the white pine once did when it was a dominant tree all along the East Coast.

If I had had more time to explore, and known where to look, I might have found the endangered greenfly orchid or the sarvis holly that grow there. What struck me most about the Black River Preserve is what impressed me when I visited remote peatlands in the Hudson Bay Lowlands, northern Alberta, Alaska, Chukotka, the Mackenzie River and Delta, and the Thomsen River Valley on Banks Island in the High Arctic. They look as if no one had ever passed through, save for the storms that have weathered them, the ground and surface water that is constantly reshaping them, and the patches of sunlight that make their dark matter suddenly glow, like the pink light of the rising and setting sun that warms up snowy mountain tops. Each place has something special—a rare orchid or moss, a songbird or animal, a butterfly or moth, a wasp or a spider or some other living thing that has yet to be discovered. Peatlands are difficult to assimilate because they are ambiguous, dangerous, and sublime, and yet, in big and small ways, they offer so much that benefits us.

Chapter 2

Central Park

Thus the inhabitants are very busy here, not only to lessen the number of trees, but even to extirpate them entirely. They are here [New Jersey] and in many other places in regard to wood, bent only on their present advantage, utterly regardless of posterity. By this means many cedar swamps are already quite destitute of cedars. . . .[1]
— Pehr Kalm, *Travels into North America*, 1750

On my many walks through Central Park, I have seen little evidence of the coniferous bogs and fens that, 13,000 years ago, rose up after a biblical flood of glacial water swept in from Lake Iroquois through the Hudson River Valley, throwing ocean circulation out of whack and cooling the climate in central North America, Atlantic Canada, and parts of Europe for a time. In today's Manhattan, sphagnum, dwarf willow, sheep laurel, Labrador tea, bog birch, high bush blueberry, and carnivorous sundews growing out of peat hummocks are either rare or no longer exist at all.

The Atlantic white cedars that once flourished along the edges of Central Park, Inwood Hill Park, and a third of the Hackensack Meadowlands are no longer there except when the tides along the Hudson River recede and reveal their decaying, twisted stumps. The Hessel's hairstreak butterfly, whose caterpillars like to feed on cedar leaves, has also long vanished.[2]

The record of Seneca, the village that was home to 225 Irish, German, and, mostly, African American men and women, some of whom had escaped slavery from places like the Great Dismal Swamp, is commemorated simply with a plaque. The village was located between

West 81st and West 89th Streets, and what would have been Seventh and Eighth Avenues, south of today's Jacqueline Kennedy Onassis Reservoir in Central Park. The houses of these people were built along the upland areas adjacent to the so-called swamp and the freshwater springs that quenched the thirst of the inhabitants.

Manhattan's flora was once at the intersection of boreal peatland species overlapping with the verdure of Virginia and the Carolinas—"peat moss and magnolias," as landscape ecologist Eric W. Sanderson describes it, the crossroads of two great traditions—the North and the South.

When plans were put in place to establish Central Park in 1856, the *New York Times* opined that the "primitive elements" of what was there then "has got to be inclosed, the marshes must be drained, bogs converted into lakes, and barren rocks covered in verdure. . . . Suitable trees must be planted, a nursery formed, roads laid out, bridges built, and lawns for recreation and military exercise created." A "system of drainage" would have to be put in place so that those "primitive elements" would have no opportunity to return.[3]

Once the people were evicted, the Atlantic white cedars were felled. What little was left of the mosses, Labrador tea, blueberry, and thorny shrubs was pulled up or plowed over, and the draining of the peaty wetlands followed to make way for the park that Frederick Olmsted and architect Calvert Vaux had designed.

In all, ten million cartloads of clay, rocks, peat, brush, and tree stumps were removed or shifted to other parts of the park area to mould gentling rolling hills or to fill in depressions left behind when the wetlands were drained.[4]

Some of the depressions were filled with water in order to create those pastoral ponds and artificial waterfalls that the *New York Times* and many others yearned for. Four million shrubs and trees were planted around them (270,000 during the first phase). Many, like ginkgo, Austrian pine, and English oak came from other parts of the world. The native conifers planted along a thirty-block stretch of road meant for horse-drawn sleighs in snow died, presumably because the chemistry of the soil did not suit them. All that is left of the American swamp is the black tupelo, whose hold in the Northeast was always tenuous, confined as it was to warm, moist lowland spots along the Hudson River–Champlain River Valley.

Central Park wasn't the garden of the Palace of Versailles, but neither did it resemble one of the varied peatland ecosystems the Dutch had seen when they arrived in 1624 and encountered the indigenous Lenape people. (The British land grants that followed were often given on the condition that wetlands of Manhattan be drained within a year's time.)

The concept of draining bogs, fens, marshes, and swamps was around long before William Byrd first came up with the idea of dewatering the Great Dismal Swamp. Centuries before, Julius Caesar had laid out an ambitious plan to divert the River Tiber through the Pontine Marshes in central Italy so that water in those wetlands could be drained to make way for farms and cities. Various Caesars and more than one pope tried and failed to accomplish this until Benito Mussolini came along and finished the job in the 1930s. For centuries, inhabitants of the Hula (Huleh) Valley in northern Israel tried to find a balance between land and water before millions of gallons of swamp water were diverted to free up 15,000 acres of peat-rich soil for cultivation in the 1950s.

It was the Dutch, though, who transformed the labor-intensive draining of peatlands into remarkable feats of hydrological engineering. For the Dutch, this work started in 1050, when the counts of Holland and Guelders and the bishops of Utrecht began offering people freedom and land-ownership opportunities so long as they made the fens and bogs in their territories habitable.

Initially, drainage amounted to little more than digging long, narrow canals that diverted water out of the wetlands into existing streams or larger rivers such as the Rhine. From the seventeenth century and onward, advances in windmill technology allowed much more water to be scooped up out of the lowlands and diverted back into the rivers beyond dikes. Not only did the Dutch turn the wetlands into farms, but they also exposed yards-thick layers of peat that could be carved up, dried out, and compressed into briquettes (*turfbriketten*). The "forgotten fuel" or "black gold," as the Irish still call it, was slow-burning and virtually smokeless if it was of good quality.

Cornelius Vermuyden introduced Dutch land-reclamation methods to England in 1630, when he was contracted to drain flood-prone Hatfield Chase in south Yorkshire and north Lincolnshire to make it "fairly fit for arable meadow or pasture." Local fenmen who hunted and fished in those wetlands rose up and attacked the Dutch laborers. Five years later, Vermuyden was commissioned once again, this time to drain the Great Fen near Cambridge. Over the course of the next decade, nine drainage ditches, ranging in distance from 2 to 21 miles, were excavated. Parliament ordered the dikes to be destroyed in 1642 in order to unleash a flood that would stop a royalist advance during England's first civil war.

The ditches and canals that Olmsted observed when he traveled to the Great Dismal Swamp in 1853 may have informed him on how to (or, more likely, how not to) drain the wetlands along the Upper West Side of Manhattan. With an abundance of free labor from enslaved people, there was no financial incentive for Dismal Swamp companies to invest in more sophisticated ways of diverting water, as George Washington wanted to do when he suggested bringing in 300 Dutchmen.

The real lesson for Olmstead came three years earlier when he and his brother John visited Birkenhead Public Park near Liverpool, England, where a more sophisticated underground drainage system was used to convert 120 acres of flat, poorly drained peatland and farmland into a park with hills, meadows, glens, walking trails, and a man-made lake. Completed in 1847, Birkenhead was the first park in Britain to be built at public expense.

The British had by this time nearly caught up with the Dutch in transforming the back-breaking practice of draining bogs, fens, and swamps with shovels and plows into a more efficient, gravity-induced hydraulic system that included stone, wood, ceramic tiles, and pipes instead of, or along with, the traditional open-ditch system.

An illiterate farmer named Joseph Elkington is largely credited for that. In the early 1760s, Elkington was so successful in helping other farmers, including the Duke of Bedford at Priestly Farm, to drain thousands of acres of bogs and fens that in 1795 the British House of Commons awarded him a gold ring and 1,000 pounds—a fortune in those days—to reveal his methods, which included the "tapping of springs," whereby groundwater is diverted away from a fen.[5] The information he provided was used to produce a Board of Agriculture book on the subject. Demand for the book was so high that it required three printings.[6]

Henry Flagg French, an American farmer who eventually became the second assistant US secretary of the treasury, promoted Elkington's technique in the United States after visiting England and Holland in 1857, the year Olmsted was hired to be superintendent of Central Park. French was obsessed not only with ways of draining wilderness wetlands for agricultural purposes but also for ways in which the common flooding of basements could be prevented. Flagg harbored the notion that the premature death of his wife had something to do with their damp cellar. (The popular French drain or "weeping tile" system that is still in use today is named after him.)

French's visit to England was inspired by John Johnston, a Scottish immigrant who used British tile and draining techniques to dewater the bogs on his farm near Geneva, New York, in the late 1820s. Fellow farmers had initially viewed Johnston as a wide-eyed lunatic for his believing that "the whole airth should be drained."[7] French and others, however, saw him as an outstanding contributor to human welfare. (New York's state education department put up a plaque that says exactly that about him.)

French saw no need to spare even the remotest wetlands that dominated much of the American landscape in his day. His vision was a kind of Manifest Destiny for getting rid of bogs, fens, and marshes. "In New England, we have determined to dry the springy hillsides, and so lengthen our seasons for labor; we have found too in the valleys and swamps, the soil which has been washed from on our mountains. . . . On the prairies of the Great West, large tracts are found just a little too wet for the best crops of corn and wheat, and this inquiry is anxiously made; how can we be rid of this surplus water?"[8]

"Elkington's method," French wrote, "cuts a drain deep into the *seat of evil*, and so lowering the water that it may be carried away below the surface." It was, he said, "a common sense remedy."

Not everyone agreed with this practice of draining swamplands. George Perkins Marsh, a Vermont farmer and well-traveled American diplomat, was horrified by the degradation that deforestation and drainage had done to European peatlands such as Tuscany's Maremma, which was so notorious as a malarial borderland of herders and bandits that it has a place in Dante Alighieri's very dark *Divine Comedy*.

"It is desirable that some large and easily accessible region of American soils should remain, as far as possible, in its primitive condition," Marsh wrote in his book *Man and Nature*.[9] Marsh was not so much against the practice of draining wetlands as he was against doing too much of it, too fast. He not only called for preservation of wetlands, but for their restoration, which was a novel concept in his time. Unlike Olmsted, Marsh believed that nature was the force, good and bad, that shaped the landscape. "Wherever (man) fails to make himself master (of nature), he can be but her slave." Learning to become a "co-worker" was the solution.[10]

There were other reasons why so many people were intent on draining peatlands. For the British, it was a way of transforming bogs into farmland, and reforesting the country with timber that, meanwhile, was being imported at great cost. Beginning in 1758, the Royal Society for the Arts (RSA) handed out gold medals to those dukes, duchesses, earls, viscounts, marquesses, bishops, and members of parliament who planted the most trees in a year. Sixty million seedlings were planted in wetlands that had been drained for that purpose.

For Olmsted, drainage provided the aesthetic that the Royal Society also strived to achieve. Olmsted's views on nature were firmly rooted in the British notion of the picturesque, which favored idyllic scenes that were pleasing, rather than sublime ones that were mysterious, daunting, or uncontrollable, as swamplands were perceived to be. Olmsted would have been appalled to hear his contemporary Henry David Thoreau say that "hope and the future for me are not in lawns and cultivated fields, not in towns and cities, but in the impervious and quaking swamps."[11]

"Without the wetland," Thoreau wrote, "the world would fall apart. The wetland feeds and holds together the skeleton of the body of nature."[12]

Public health was also paramount in the mind of Egbert Viele, the state engineer for New Jersey who became chief engineer for the

Central Park project. In the early 1850s, when there were more than three million people rubbing elbows in Manhattan, there were few places where one could go for a walk in order to enjoy some fresh air. Heavy rain and spring water that fed bogs often overflowed into city streets, where waste was typically disposed of in backyard outhouses or into the gutters, or dispatched back into those bogs and streams. In 1854, the runaway best-selling book *Hot Corn* advised people who planned to venture out to some of the poorer parts of the city to "saturate your handkerchief with camphor, so that you can endure the horrid stench."[13]

But it wasn't just human and animal waste overflowing in the streets that concerned New Yorkers. Viele and others also believed that organic soil and damp peat in New York City's coniferous swamps spread poisonous vapors into the air just as the city's stinky sewage did. Viele said as much when he expressed the view that good "sewerage" in a future Central Park was not enough to eliminate that threat of miasma. Unless all of the swamps were drained, he wrote in the *New York Times*, the park would "remain what it is now, a pestilential spot, where rank vegetation and miasmatic odors taint every breath."[14]

The idea that wetlands can be dangerous to public health was put forth by the Greek physician Galen (AD 129–216), who convinced the Romans he was attending to that the Antonine Plague wasn't God's punishment for opening tombs and ransacking the city of Saleucia, but rather poisonous vapors rising from marshes where uninterred bodies of warriors were rotting. At its peak, the plague was killing as many as 2,000 Romans each day.

This noxious miasma was also thought to behind the bad air (*malaria* in Italian) that halted Hannibal's march on Rome, the Roman's own attempt to conquer Caledonia, and the British failure to seize Haiti in the 1790s.[15] It was the invisible menace behind the Black Death that killed fifty million people in the mid-1300s, as well as the cholera pandemic that spread across the world from its original reservoir in the peaty depths of the Ganges Delta in 1817.

In the seventeenth century, the English physician Thomas Sydenham, the "father of English medicine," introduced the notion of an "epidemic constitution of the atmosphere"—a miasma that had its origins, he speculated, in the "bowels of the earth." This medical miasma theory gained scientific credibility when it was championed by Max Joseph von Pettenkofer (1818–1901), the Bavarian chemist who proclaimed the need for clean air, water, and personal hygiene as the three pillars of public health.

English social reformer and sanitary engineering proponent Edwin Chadwick attested to its credibility as well. He was so convinced that stinky decaying matter such as that which lay at the bottom of a bog was responsible for disease that he told a parliamentary committee something that must have taken some time to sink into their ever-so-proper sensibilities. "All smell is, if it be intense, immediate acute disease; and eventually we may say that, by depressing the system and rendering it susceptible to the action of other causes, all smell is disease."[16]

It's easy to understand, then, why the public had a phobia about the misty marshes and foul-smelling sulfurous bogs that surrounded them. Most no longer believed that God was sending disease to punish them for their sins. They recognized that it was something external that created an imbalance of the "four humors"—blood, phlegm, yellow bile, and black bile—which were connected to the four seasons.

Physicians of the nineteenth century often associated chills and fevers with climate and the natural environment, using terms such as *country fever*, *swamp fever*, and *marsh fever*. Cholera was one of those truly horrifying diseases whose origins were thought to be miasmatic, and most everyone knew of someone who had been stricken with cholera, because it was so frightfully common. A victim could be healthy at the beginning of the day and dead by dinnertime.

Medical miasma was such a well-accepted theory that it was applied to almost any illness that had even the remotest connection to anything odorous. In 1844, a British doctor may have been trying to be kind when he attributed a butcher's wife's extreme obesity to her inhaling the odor of beef.[17] (But then again, probably not.)

All manner of newspapers, magazines, and farmer's almanacs ascribed to the miasma theory. Editors and publishers sensationalized miasma's deadly reach for a gullible public that had been raised on grim Gothic tales and superstitions that made them close their eyes, plug their noses, and cover their ears when they passed by a bog at night.

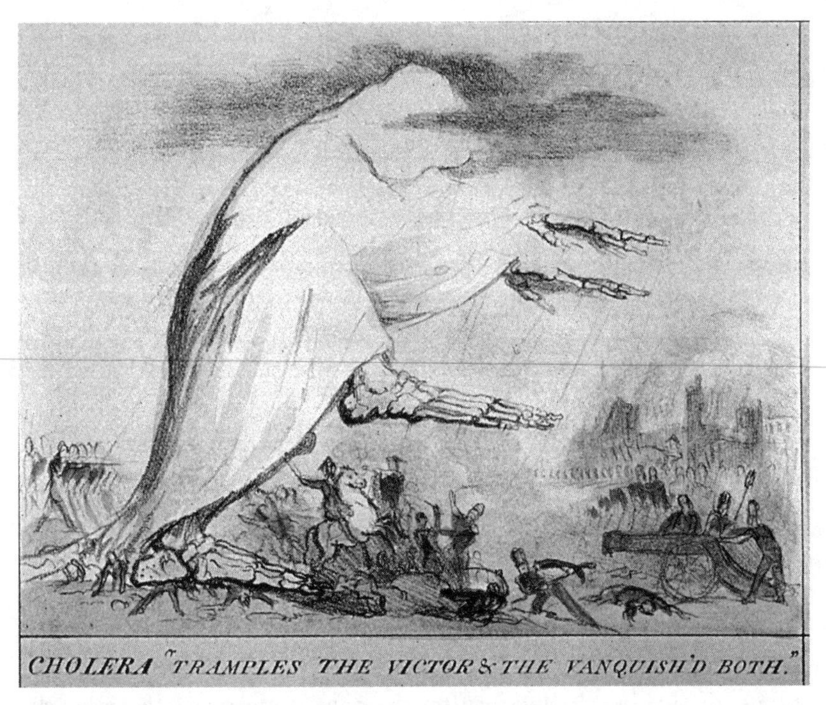

Cholera, depicted as a large, shrouded specter with skeletal hands and feet, indiscriminately crushes soldiers on both sides of the battlefield in this illustration by artist Robert Seymour, October 1831. (Photo: National Library of Medicine.)

In an 1846 issue of the popular satirical magazine *Punch*, a seriously worded editorial described how gases "ascending into the atmosphere, will mingle with it, and contaminate it for miles around." An accompanying poem, "The Vampyre," evokes frightening images of "bloodshot moons" and "a glimmering vapour creeping over the ground . . . the reek of the charnel, the poison of death."[18]

Farmer's almanacs were obsessed with cures without knowing the cause of a miasmatic event, other than it originated in bogs and moist landscapes. In 1823, the *New England Farmer* offered false hope to swampland owners with news that the planting of sunflowers would purify the air because their leaves "have the property of absorbing the miasma."[19] *Hostetter's United States Almanac of the Use of Merchants, Mechanics and Planters and All Families* sold their trademark "Stomach

Bitters" as a means of defying "the miasma of unhealthy soils, the poison of unwholesome water. . ."[20]

For city dwellers, the best way of avoiding a miasmatic attack was to wear a nosegay or "tussie-mussie." It was a scented flower attached to the lapel. It was thought to work as well as colognes, handkerchiefs, and even cigars in warding off these airborne diseases. Not to be left out, the Italians marketed a spiced, aromatic wine, Barolo Chinato of Piedmont, which was traditionally drunk as a digestive until someone came up with the idea that it could ward off the "bad air" that carried malaria.

You didn't have to be uneducated or gullible to buy into any of this. Mary Shelley, the author of *Frankenstein*, was a late convert to the miasma theory after aligning herself initially to "contagion," which held the view that close contact and bodily fluid, rather than noxious miasma, spread disease. She had lost family members to these so-called airborne diseases. In her third novel, *The Last Man (*1826), a form of noxious miasma is responsible for a plague that sweeps across the world at the end of the twenty-first century, much like the coronavirus did nearly two hundred years after the novel appeared. "We feared the coming summer. Nations, bordering on the already infected countries, began to enter upon serious plans for the better keeping out of the enemy. We, a commercial people, were obliged to bring such schemes under consideration; and the question of contagion became matter of earnest disquisition."[21]

In the United States, Edgar Allan Poe left New York for Virginia in 1849 when cholera was sweeping through the city. In his story "The Fall of the House of Usher" (1839), the miasma that is responsible for so much death is "an atmosphere which had no affinity with the air of heaven, but which had reeked up from the decayed trees, and the gray wall, and the silent tarn—a pestilent and mystic vapor, dull, sluggish, faintly discernible and leaden-hued."[22]

Poe, we know, was fascinated by the Great Dismal Swamp, which was thought to be a vector for disease. In his much maligned and misunderstood poem "Dream-Land" (1844), ghouls and ghosts haunt a cold and terrifying landscape such as the Great Dismal was perceived to be. While some literary critics insist that much of the gothic horror that Poe depicted in other stories such as "The Masque of the Red Death" (1842) was more a metaphor for the human state of mind than for physical illness. No one reading his books in Poe's own time noticed the difference. As Harold Bloom later noted, metaphor or not, it was an ingenious way of "exposing our common nightmares and hysteria lurking beneath our structured lives."[23]

Like will-o'-the wisps, Jack-o'-lanterns and Cajun fairy superstitions, the miasma theory was wired into the human brain, enduring long after it was disproved. When British physician John Snow traced an 1854 outbreak of cholera to polluted wells in London, even John Simon, the chief medical officer of London, had his doubts, as did Dr. William Farr, who believed damp soil at low elevations, especially near the banks of the River Thames, contained organic compounds which produces *miasmata* in the atmosphere. Farr was no quack. He was a member of the Scientific Committee for Scientific Enquiries in Relation to the Cholera Epidemic of 1854. "After careful inquiry, we see no reason to adopt this belief," he wrote after reviewing Snow's report. "We do not feel it established that the water was contaminated in the manner alleged; nor is there before us any sufficient evidence to show whether inhabitants of that district, drinking from that well, suffered in proportion more than other inhabitants of the district who drank from other sources."[24] Farr regarded polluted water as just another source of miasma. It wasn't the drinking of the water that was the problem, it was the amount of organic material entering the air from the evaporation of polluted water.[25]

The New York Sanitary Association didn't do much to dispel fears but rather exacerbated them when one of its members proclaimed in 1859 that "it is an undisputed fact that miasmatic influences are a prominent causes of diseases. And it is a recognized necessity for a healthful condition of the country that water would be prevented from accumulating where it would be stagnant. Water thus held becomes sour, and passes off in the shape of mists. There is hardly any region in the country which does not require a greater or less amount of drainage."[26]

Henry Flagg French exploited these fears with the passion of a religious zealot imploring the public to hear the words of his prophecy. "From the time when Noah and his family anxiously watched the subsiding of the waters into their appropriate channels, to the present," he wrote in the *New England Farmer* in 1859, "men must have felt the ill effects of too much water, and adopted means more or less effective, to remove it." French confidently claimed that "wet, swampy districts of the country are usually afflicted with agues and fevers, and other forms of disease, from which dry regions are exempt."[27]

French was considered to be a visionary. Founders of the Massachusetts Agricultural College (now the University of Massachusetts, Amherst) made him their first president. His books on the subject of draining swamps found a large, receptive audience, as did many other books on the subject. French's best-selling, 381-page *Farm Drainage* of 1857 followed the success of Henry Stephens's *Book of Farm* in 1847. George Waring's *Draining for Profit, Draining for Health* in 1867 preceded another by civil engineer Allen Boyer McDaniel in 1879. McDaniel agreed with what the others had preached. "A large swampy piece of land is a blot on the landscape, a source of ill health, and perhaps a calamity to the people of adjacent communities."[28]

Not surprisingly, lawmakers got caught up in the hysteria and the pressing need to convert wetlands into farms and safe havens. Swamp Land Acts were passed by Congress in 1849, 1850, and 1860. The first version of the act was designed primarily to be a mechanism for turning over federally owned wetlands to Louisiana, so long as the state agreed to drain the land for the development of farms and other productive uses. Subsequent versions of the act, however, included Arkansas, Alabama, California, Florida, Illinois, Indiana, Iowa, Michigan, Mississippi, Missouri, Ohio, and Wisconsin, and then Minnesota and Oregon. All told, nearly 65 million acres of so-called swampland—bogs, fens, and marshes—were transferred to 15 states on the condition that they drain these wetlands. In January 1850, a judiciary committee for the state assembly of California made it clear that peatlands such as the Sacramento–San Joaquin Delta needed to be drained because they bred disease and inhibited the expansion of agriculture.

None of the newspapers back then second-guessed the wisdom of Congress giving away so much federal land. Most cheered the transfers of wetlands. "As fitted for human habitation, the American continent was not adapted to nurse a race of bog-trotters," the *New York Times* opined in 1865.[29]

Many of the targeted wetlands, such as the Sacramento–San Joaquin Delta, the Everglades in Florida, and the Great Black Swamp in Ohio, lower Michigan, and northern Indiana, were among the biggest reservoirs of peat in the country. It didn't matter that there were indigenous people like the Seminoles of the Everglades, the Ottawa of the Black Swamp, or the Wintun, Maidu, Miwok, and Yokut tribes of California living and hunting in these places, just as successfully as the Powhatans and maroons had in and around the Great Dismal Swamp. Like the Powhatans, most of these people were driven out of their territory by scorched-earth policies by which the army was brought in to force them out. Only the Seminoles successfully resisted efforts to terminate a lifestyle in a peatland ecosystem shaped by constantly fluctuating water levels. But they did so only after three miserable wars.

To meet this increasing demand to drain the country's swamps, marshes, bogs, and fens, drainage tile factories shot up everywhere, from the tidal marshland regions of New Jersey and New York State to the boreal peatland forests of Ohio, Michigan, Illinois, Wisconsin, and Indiana, where, in the 1800s, half of the surface area in the northwest part of the state was covered in water for at least half the year. By one estimate, in 1884 there were more than 1,140 drainage tile factories in eight states that recorded the numbers.[30] By another estimate, there were more miles of drainage ditches than of highways by middle of the twentieth century.[31]

Even states that did not directly benefit from the Swamp Drainage Act took it upon themselves to rid themselves of disease-spreading wetlands. By the end of the First World War, the state of New Jersey had trenched approximately 120,000 acres of salt marsh. It involved the cutting of 18,244,217 linear feet of open ditches, 10 inches wide and 24–30 inches deep, in order to destroy mosquito-breeding habitat.[32]

There was still a lot of peat in the United States at the turn of the last century when the US Geological Survey crudely estimated that 14 trillion tons of it could be harvested for fuel and other purposes in the Lower 48.[33] What we see in many of those landscapes today bears almost no resemblance to what they looked like 200 years ago. Ohio has lost 90 percent of its peatlands, second only to California. Parts of the Sacramento–San Joaquin Delta now lie 26 feet below sea level because so much peat was dug up, dried out, plowed over, and compressed by heavy farm machinery. The peatlands of Orange County in California are now better known for orchards than for the wild orchids that once grew there.

The Everglades has a lot less diversity (180 species on the state's threatened list) since water was drawn from Lake Okeechobee in 1884 as an initial step to drain them. Greed, incompetence, and a gross misunderstanding of the role that climate, storm surges, and freshwater hydrology accounted for scheme's failure. There was just too much rain, and too much water surging in from the ocean on one side and from rivers, lakes, and springs flowing in from other directions. Draining the Everglades was akin to stopping the flow of a slow-moving river, 100 miles long and 60 miles wide. Not even the Dutch could have done it had they been given a chance.

None of this stopped unscrupulous land speculators like "Dickey" Bolles from selling peatlands in the Everglades in the early twentieth century to hundreds of people who had no idea that the plots they bought were still under water. The Florida governor, his predecessors, and US Secretary of Agriculture James Wilson were in on it as well. "Doubting Thomases who were waiting for the Everglades to develop before buying would regret it all their lives," Wilson warned.

Over time, it was left to Congress and the US Army Corps of Engineers to deal with the flooding and the fires that came with the withering that followed drainage. Instead of removing the dikes and stopping the flow of people moving in, Congress in 1930 authorized the Corps to construct 67.8 miles of levees along the south shore of Lake Okeechobee and 15.7 miles along its north shore.

Some scientists saw nothing wrong with these attempts to drain peatlands like the Everglades. When Frederick Olmsted's son led an effort to turn part of the Everglades into a national park in the 1930s, William T. Hornaday, one of the country's most respected zoologists, objected. "A swamp is a swamp," he scoffed. There is "mighty little that was of special interest, and absolutely nothing that was picturesque or beautiful," he said about his visit.[34] He allowed that the Everglades were less "repugnant" than most swamps, but still not worthy of national park status.

Hornaday's views did not prevail, thanks to newspapers such as the

New York Times and magazines like *National Geographic* that came around to the idea that the Everglades might be worth preserving. But making the Everglades a national park did not put an end to threats to the ecosystem.

Since the park came into being in 1934, there have been lawsuits to limit the amount of pollution from farms, to stop mining in the historic unprotected part of the Everglades, and to prevent the US Army Corps of Engineers from dredging to create a site for a biotech facility in western Palm Beach County. A lawsuit by landowners sought $50 million in compensation from the government for the US Army Corps of Engineers' turning the once-clear St. Luce River into a waterway that could no longer support fish, oysters, clams, pelicans, osprey, and wild ducks. The Seminoles hired lawyers to get their share of justice. The US Air Force was forced to back away from a plan to build an airport that would have further altered water flows in the Everglades.

The craziest scheme of all was a 1960s plan for a jetport that would have covered 39 square miles of the Great Cypress Swamp, which supplies 38 percent of the water flowing into the Everglades National Park. It would have been as large as New York's JFK, Chicago's O'Hare, and the Los Angeles and San Francisco airports combined. There were to have been twelve terminals and eight three-mile-long runways.

A great deal of work—at a cost, by one estimate, of $8 billion—has been done to restore the Everglades to some semblance of what it once was. Progress has been made on a number of fronts. But the ecological damage done over the past 150 years is staggering. Half of the 4,200 square miles of wetland that was there in the 1800s no longer exists. Three-quarters of the peat has been lost—and with the loss of that peat, at least 1.3 billion metric tons of carbon dioxide has been released into the atmosphere. The removal of peat has resulted in the land subsiding by as much as five feet.[35]

Hardest hit are the floating islands of peat that so fascinated botanist William Bartram when he explored the Everglades in the 1770s and harvested seeds that he sent to collectors in Great Britain. Known also as tree islands or pop-ups, these peat islands can range in size from a few square feet to hundreds of acres. The larger floating islands are buoyant enough for trees to grow up to fifty feet. They are critical to the survival of most species in the Everglades because they offer safe nesting and denning sites, and they absorb most of the nutrients that flow in from nearby farms. Rising sea levels, floods, fire, and invasive species are sinking hundreds of them.

I had thought that the crazy chapters in the history of the Everglades had come to an end when President Bill Clinton endorsed a 1996 plan that allocated hundreds of millions of dollars toward restoration. But then in 2018, I met Craig Fugate, a former FEMA director and former director of the Florida Division of Emergency Management. Fugate had managed a countless number of emergency responses to hurricanes in Florida. We were both speaking at a conference in Sacramento when he told me of a plan to create the world's biggest mall on a wetland that drains into the Everglades. In addition to shops and restaurants, the $4-billion project was even more bizarre than the theme park I had visited along the Pocahontas Trail in Virginia. Instead of Rhine River cruises and amusement parks named after European countries, this one in Florida would have an indoor ski slope, along with submarine rides, a water park, and a skating rink. When I did a little more investigating to see who the developers were, I discovered that they were the same ones who built the American Dream Mall on a swampland that had been drained in the Meadowlands of New Jersey before it opened in 2019.

Chapter 3

Peat and Endangered Species

About a mile south of the small town of Columbia on the soggy Albemarle Peninsula of eastern North Carolina, there is a fenced-in, forested enclosure that was constructed to hold red wolves, which have been periodically reintroduced to the Alligator River and Pocosin Lakes National Wildlife Refuges. Howard Phillips, the US Fish and Wildlife Service refuge manager for Pocosin Lakes, brought me there to give me an update on how the effort to save one of world's most endangered animals was going.

Not well by any measure. In 2005, there were as many as 150 red wolves that had been successfully reintroduced to the region. When I was there with Phillips fifteen years later, there were just twenty. No pups had been born in the wilderness the spring before. No pups were born the year I was there. That has not happened since the reintroduction program began in 1987, when four breeding pairs were released into Alligator River Refuge about ten miles east of where we were that day.

"The wolf reintroduction program is in serious trouble," Phillips told me as we passed through two locked gates that are designed to keep the wolves in, and people with malicious intentions out. The possibility that someone opposed to the reintroduction might toss poisoned meat over the fence is not out of the question, he allowed.

"They're getting hit by cars or being shot by hunters mistaking them for coyotes. Some people are shooting them simply because they don't want them on their land."

Pocosin Lakes and Alligator River are two of fifteen national wildlife refuges in North Carolina that were once the haunt of red wolves and panthers and part of a statewide, heavily fragmented peatland ecosystem that continues to provide refuge for rare species such as the red wolf, the mountain crayfish, the Venus flytrap, and the Southern Appalachian purple pitcherplant, and reptiles like the tiny bog turtle that digs its nests in clumps of sphagnum moss.

There were two wolves in the enclosure late that afternoon. They were healthy animals that have the striking looks of both wolves and coyotes: a short, tawny-to-rust-colored coat, big ears with cinnamon-colored tips, ridiculously long, skinny legs, and a fixed, impassive, amber-eyed stare. They are almost twice the size of coyotes, and about two-thirds as big as the gray wolf. They reminded me of the smallish eastern timber wolves I have seen on canoe trips in Algonquin Provincial Park in Canada.

Contrary to what critics of the reintroduction program have stated publicly, and scientists have debated almost endlessly, red wolves are a distinct species, not a hybrid produced by wolves mating with the coyotes that have recently moved into the region. An expert panel, appointed by the National Academies of Sciences, finally made that conclusion in 2019.

I watched with keen interest that day as one wolf paced back and forth behind a mounded man-made den, stopping and craning its neck at times to see what we were up to. I strained at times to see the other wolf that showed itself only once. The sound of cars and trucks roaring by on the nearby highway must have been stressful for highly territorial animals like these that rely so much on their hearing and other senses.

I try not to be anthropomorphic, but I found this to be a profoundly melancholy scene, given the background to the story. Neither one of the wolves will ever be returned to the wild. They are there as exhibits

A red wolf attempts to evade capture by biologists checking on its health and well-being. Routine checkups were once a regular practice at the Alligator National Wildlife Refuge before the reintroduction program was put on hold. (Photo: Steve Hillebrand, USFWS.)

only, intended to win over people who know nothing about the plight of red wolves since they were declared to be functionally extinct in 1980, as well as those who might be undecided about the wisdom of reintroducing this species back into the wild.

The reintroduction program had been on hold since 2014 because of a small group of landowners and conservative politicians who want to be rid of the animals. In 2015, the North Carolina Resource Wildlife Commission, the state agency responsible for conserving and sustaining fish and wildlife resources through research and scientific management, backed the critics by formally requesting that the US Fish and Wildlife Service declare the red wolf extinct in the wild and terminate the reintroduction program.

By the time I arrived, Fish and Wildlife officials in the Atlantic office had already given in to some of these demands. Instead of aiming for a population of 200 animals in five counties, as was originally planned, they were proposing to hold the numbers to approximately twenty in the Alligator National Wildlife refuge and the adjoining bombing range. They had also given landowners permission to kill wolves if they showed up on their property.

What the future holds was anything but clear. In November 2018, the Southern Environmental Law Center, on behalf of conservation groups, secured a court order declaring that the agency violated the law in gutting protections for red wolves. The court also made permanent a previous (September 2016) order stopping the Service from capturing and killing, and authorizing private landowners to capture and kill wild red wolves not posing a threat to human safety or property.

The political situation got murkier in November 2019, when North Carolina's governor, Roy Cooper, weighed in and sided with conservationists who had gone to court to force the Service to get on with a recovery plan. In the winter of 2021, a federal district court judge ruled in their favor, agreeing that the Fish and Wildlife Service was in violation of it duty to conserve endangered species. The judge ordered the Service to draw up a plan to release more wolves into the wild.

Like the mountain bog turtle of the southern Appalachians and the southern bog lemming in the Great Dismal Swamp, red wolves are relics of the Last Glacial Period, when glaciers came close but did not cover the Albemarle region. Sea levels were so much lower 12,000 years ago that Albemarle Sound, Pamlico Sound, and Chesapeake Bay did not exist as they do today. Fossil evidence suggest that the red wolf shared the tundra-like environment with mastodons, mammoths, bison, seventeen-foot-long sloths, and other animals, many of which were driven to extinction by a warming climate that led to sea ice melting, glaciers retreating, sea levels rising, and rivers like the Roanoke, the largest drainage system in the North Carolina, getting inundated with ocean water.

Survivors that they are, red wolves haven't gotten anywhere near the attention that gray wolves have received even though the reintroduction program started seven years before grays from Canada were released into Yellowstone National Park with international fanfare. It's a remarkable irony, because the red wolf is a quintessentially American animal. Unlike the gray wolf, or the emblematic bald eagle, the red wolf is not found anywhere outside the United States.

Talking to conservation leaders and red wolf advocates like Ron Sutherland, chief scientist of the Wildlands Network, it's easy to understand why they blame the US Fish and Wildlife Service for failing to turn this into a good-news story, as the US National Parks Service did in Yellowstone. Instead of touting this as the first successful reintroduction of a functionally extinct animal back into the American wild, the Service has allowed misconceptions about red wolves to undermine reintroduction efforts.

There is no evidence that wolves are decimating deer populations, as some people think. That's pretty obvious to anyone who spends a day in the Pocosin Lakes or the Alligator refuge. The Fish and Wildlife Service never promised that wolves would not venture onto private land—as they often do. Contrary to what critics claim, Sutherland points out, the public supports the reintroduction with numbers that would make even the most risk-averse bureaucrat giddy. Only 30 of 108,124 people and organizations expressed a desire to end the reintroduction program when the Fish and Wildlife Service called for public comment in 2018. More than 99 percent were in favor of doing more to save the red wolf. This exceptionally good news was revealed not by the US Fish and Wildlife Service but by Wildlands Network and partner organizations that worked together to tabulate the results.

The biggest misconception about red wolves is that they are not native to North Carolina.

In fact, red wolves are native not just to North Carolina, but to a big chunk of the United States. The animals once roamed freely from central Texas eastward to Virginia and the Carolinas, south to the Gulf of Mexico and north to central Missouri and southern Illinois. For a time, they extended their reach into New Jersey and Manhattan, where, in the 1720s, the last wolves were systematically removed from the cedar forest and peaty marshes of what is now Inwood Hill Park.

Red wolves were present in such large numbers when the settlers arrived in Jamestown in 1607 that bounties were placed on them. Some of the colonists called them "swamp wolves"—demonic animals sent by God Almighty to punish those who had sinned or forsaken Him.

(Wolf Swamp is the name of a river in Virginia.) The Jamestown colonists were haunted for decades by the sight of one of their dead being torn up by a pack of wolves during a famine in 1609.

There was more to it than that. The colonists came from Europe, where wolves had been demonized since at least 1281, when King Edward I ordered the extermination of all wolves in England. By the time the settlers first set foot in the New World, wolves were extinct in England and only a few packs remained in remote peatlands like the Flow Country of Scotland.

One of the very few colonists who had anything remotely good to say about red wolves was William Byrd, the womanizing plantation owner who helped define the boundary line between Virginia and North Carolina in 1728. Camped out in the Great Dismal Swamp one night, Byrd wrote almost wistfully about red wolves howling "with the yell of a whole family of wolves, in which we could distinguish the treble, tenor, and bass, very clearly."

Byrd allowed that the wolf "will not attack a man in the keenest of his hunger, but run away from him, as from an animal more mischievous than himself." And yet, he observed, "the inhabitants hereabouts take the trouble to dig abundance of wolf pits, so deep and perpendicular that when a wolf is tempted into them, he can no more scramble out again than a husband who has taken the leap can scramble out of matrimony."[1]

Moravian Bishop August Gottlieb Spangenberg made a similar observation when he traveled from the coast of North Carolina to the interior of the colony in 1752. "The wolves here give us music every morning, from six cornets at once, such music as I have never heard in my life." They are "not like the wolves of Germany, Poland, and Livonia," he wrote, "but are afraid of men, and do not usually approach near them." Yet the colonial government, he noted, paid a bounty of ten shillings for each wolf killed.

Bounties were not enough to eradicate wolves, which reproduce prodigiously so long as there is food for the pups. Enlisting the help of indigenous people to rid the country of animals was seen as necessary both to protect livestock and to turn Native Americans into Christians. Raising livestock was viewed as part of the English way of life, going hand in hand with Christianity and the notion that the "English way" was superior to any other.[2] It was also, in all probability, a way of distracting native people from attacking them as they routinely did in North Carolina right up until the Tuscarora War of 1711 to 1715 which was among the bloodiest Indian wars in early American history.

Beginning in 1656, native chiefs in Virginia were offered a cow for the heads of eight wolves brought to the county commissioner. (Colonists were offered much more in the form of tobacco for one wolf.) A similar offer in the Albemarle Peninsula region "made the Indians so active," according to one observer at the time, "that they brought in such vast quantities of their heads, that in a short period of time it became too burdensome to the country, so that it now laid aside."[3] African Americans, both slaves and free men, were eventually enlisted in the cause. One man, named Peter, whose mother was a native North American and father an African, earned his freedom from his Quaker owner for "meritorious actions in destroying vermin such as bears and wolves." In New York, an act passed in 1775 stipulated that a bounty to be paid for wolves killed by slaves must go to the master.[4]

The systematic killing of red wolves and other predators expanded as the fur trade with Native Americans declined and as succeeding generations became increasingly intent on draining bogs, fens, and swamps, cutting down peatland forests and turning them into farms that produced tobacco, corn, and cotton, as well as livestock. Pigs, cows, and horses were easy prey because they were allowed to roam freely in forests occupied by wolves. (The penning of livestock was uncommon until the eighteenth century.)[5]

On and on the killings went, sometimes in the most labor-intensive manner. In 1760, 200 men hiked in from all directions into a 700-square-mile area of central Pennsylvania where red wolf habitat converged with that of the gray wolf.[6] Included in the list of animals killed, there were 109 wolves, 41 cougars, 114 bobcats, 112 foxes, 18 bears, and 12 wolverines.[7]

Still, it was not enough to rid the country of red wolves. As late as February 1768, records from the Tyrell County courthouse on the north side of the Albemarle Peninsula demonstrated that wolfers were still engaged in a brisk business, bringing in heads of wolves and wild cats to claim a bounty.[8] In 1900, the bounty system in some southern states was paying $20 for each dead wolf, a reward that continued to be doled out up until the 1960s, when there were too few red wolves to make the hunt worthwhile.

The last of the red wolves were found on the coastal peatlands along the Texas–Louisiana border, where few hunters were inclined to go and where the landscape was too wet and salty to be drained for agricultural purposes.

No one knew how many wolves there were until Curtis J. Carley, a US Fish and Wildlife Service biologist, went in with aerial infrared NASA photos, police sirens, and modified leg traps to find and catch them. It soon became obvious that these reluctant swamp dwellers were not faring well in habitat filled with razor-edged sawgrass, ticks, lice, and biting flies. Much of the animals' fur was shaved off by sawgrass or balded by mange. Their hearts and lungs were so badly infected that some wolves fell over and died of stress when they were captured and placed in pens.

By the time Carley and his colleagues sorted out the coyote–wolf hybrids from the pure wolves, they ended up with just seventeen reds. In 1975, everyone realized that this small number was not enough to sustain a recovery. In order to save the red wolf, the last of the wild wolves would have to be captured and placed in a captive breeding program.

At first glance, the Albemarle Peninsula didn't strike me as wolf country. The land here is dominated by brackish marshes, coastal plains, Carolina bays, remnant clusters of Atlantic white cedar, mixed pine–hardwood forests, pond pine, savannah, and ghost forests—naked stands of trees that have been killed by saltwater intrusion brought on by rapidly rising sea levels and increasingly powerful tropical storm surges.

What little dry land there is, is dissected by tea-colored tributaries branching out from inland rivers like the Roanoke and Scuppernong, the Algonquin word for the big white muscadine grapes that grow in the wetlands. Alligators up to twelve feet long do live in the Alligator River, along with sharks, dolphins, and seals that feed along its mouth.

The peninsula is home to a naval bombing range, several industrial farms, and many small ones. There are more dirt and gravel roads than paved ones, and many zanily named places such as Lizard Lick, Gum Neck, Nags Head, Duck, and Buffalo City, a ghost town that once prospered with tree felling and moonshine enterprises.

Some people joke that there are 5,000 snakes in the world and 4,998 live in North Carolina's swamps, and that there are 10,000 spiders, every one of which lives there as well. If it grows, they say, it will stick to you. If it crawls, it will bite you. Most Tar Heels are tired of hearing these jokes. But one person I talked to in a bar one night while eating fried shrimp and cheese grits did allow that an old-timer once famously stated that the pocosins are so thick and thorny that his hound dog had to back up to bark.

It was wet and windy when Howard Phillips and I set off from the refuge headquarters in Columbia earlier that day. The storm clouds were sailing across the sky with a speed that made me dizzy looking up at them. I wondered whether my rain gear would hold up. Seeing that I had not brought along a lunch, Phillips stopped by a gas station. "Outside of the tavern down the road, which is closed, there's not much for takeout around here," he said.

Phillips bought an armful of ready-made sandwiches that were being kept warm under lights. I opted for several bags of roasted nuts and a chocolate bar.

Driving along the northeast corner of the refuge, Phillips stopped to give me a view of Phelps Lake, which is an anomaly in the Pocosins because it is crystal clear, not tea-colored like most water bodies in the peatlands of the peninsula. "If you're looking for a place for largemouth bass fishing, this is it," he said almost wistfully. "It's the best fishing in the state."

In 1985, when the water levels dropped because of a severe drought, fishermen spotted dugout canoes at the bottom of the lake. Archeologists came in to have a look and found 30 cedar canoes and thousands of artifacts that date back as far as 2430 BC. One of the canoes was 37 feet long, more than twice the size of a typical canoe that most people paddle today. They were relics of indigenous cultures that are now long gone.

When the Tuscarora War ended in 1715, native Americans who did not migrate west were given a four-square-mile reservation adjacent to Lake Mattamuskeet. A small group of about twenty subsistence families were living there in 1731. In 1792, their reservation lands were sold, and they ceased to exist as a distinct culture on the Albemarle.

Phelps Lake is one of number of elliptically shaped depressions found along the coastal plains of the Atlantic. Mysteriously, they are all oriented in the same direction, and are known as "Carolina bays" in Georgia and the Carolinas, "Maryland basins" in Maryland, and "Delmarva bays" in Delaware. They can be lakes, like Phelps Lake. Or they can be sphagnum wetlands that accumulate peat.

The debate over how these depressions were formed is far from settled. One far-fetched theory holds that they were once whale wallows, where cetaceans thrashed about on the bottom of the ocean before sea levels dropped and exposed the depressions. Others insist, also without much evidence, that they were created by a shower of meteorites. A more likely explanation is that wind and wave action or some geological force molded them into these shapes.

According to Native American legend, Lake Mattamuskeet was at one time made waterless following years of drought. (*Mattamuskeet* is the Algonquin word for "dry dust.") Ceremonial fires were built to encourage the rain gods to come. The gods, however, were displeased with what was being offered, so instead of rain they sent a giant firebird to fan the flames with its enormous wings. The fire burned far and wide, and deep into the ground, for a period of thirteen moons. Wildlife fled the area.

Facing starvation, a young maiden prayed and asked the gods for mercy. Her prayers were answered with a deluge that filled the depressions left behind by the fire that had burned deep into the ground. When the tribes returned, they found a huge lake, now fourteen miles long and five miles wide, teeming with fish and waterfowl.[9]

Fantastical as this legend would seem to be to those like me who are visitors or recent transplants, scientists found some truth to it when they drilled into the lake bottom and extracted cores that contained charcoal mixed with peat and estuarine salt. It was clear evidence of past fires that burned forests in at least three of the bogs in the region.[10]

In the Carolinas, the wetland depressions are home to many carnivorous plants such as sundew and the Venus flytrap, plants so beguiling that Charles Darwin opined, "I care more for *Drosera* (sundew) than the origin of the species." He described it as a "wonderful plant or rather a most sagacious animal. I will stick up for *Drosera* to the death of my day."[11] Like sundews that I have seen as far north as the Arctic Circle, Venus flytraps can't get enough of the food they need from the peaty, nutrient-deprived soils, but they've evolved to trap and consume insects. The flytraps do this by counting the pressure points produced by any force that touches them. A single touch is not enough for them to expend the energy to snap shut. Two or three pressure points signals that something juicy is waiting to be eaten, which, once they've trapped their prey, they do with enzymes that break down the flesh of the insect struggling in what Darwin called "a horrible prison." No other carnivorous plant moves as fast as the Venus flytrap.

Another carnivorous plant—the various pitcher plants that thrive in the peatlands here and in the southern Appalachian mountain bogs—was "so different from all known plants" when John Tradescant the Younger plucked one from the Great Dismal Swamp in 1637 that it would require a century of horticultural effort to get it to bloom in a garden, as British botanist Peter Collinson finally did when he planted it in an artificial bog.[12] He got the specimen from American botanist John Bartram, who spent a lifetime in swamplands collecting seeds from unusual plants.

Phelps is a very shallow, 16,600-acre lake that was landlocked until the eighteenth century, when plantation owners began cutting canals to drain bog water into the Scuppernong River. The impetus to do so came in 1794, when the state's general assembly literally gave away swamplands. Land speculators were quick to buy up five million acres on the cheap. Enslaved laborers were then brought in to dig ditches that were up 20 feet wide and 6 feet deep.

It seemed like a good idea when Edmund Ruffin (1794–1865), a highly respected agriculturalist (but better known in his day as an advocate for slavery, states' rights, and secession), came in to assess the results. Ruffin had "never seen such magnificent growth of corn, upon such large spaces" as in the drained areas. "Proper drainage alone would double the productive value," he wrote, "and the profit of the whole great area of what is usually considered the *now dry* land, and of the firm and partially drained swamps."[13]

Buoyed by these early successes, drainage projects expanded to other parts of the Pocosin, such as Pungo Lake and Lake Mattamuskeet. The most ambitious was one funded by the state Literary Board, which intended to use the profits to promote better education.

It didn't work out as well as hoped. Ruffin realized this when he saw how many cornfields failed after just two or three years of producing fine crops. It wasn't that the land was "worn out," as some people thought, Ruffin noted. It had more to do with subsidence that occurred when the peat was exposed and oxidized. Water rushed into the depressions, leaving roots to rot.

Hurricanes and tropical storms, still common in the Carolinas, made farming that much more difficult. By 1842, land around Phelps Lake was being sold for rock-bottom prices. Many farmers thought they would not have enough food to carry them through the winter. "We have reason to expect a most desperate sickly fall & I fear a deadly one & God knows who will be able to live," one farmer noted. "Such a storm & such destruction I have never before seen. Everything blew to pieces."[14]

As we traveled through the shrubby pocosins, I began to appreciate why the Albemarle Peninsula was the most logical place for red wolves to be reintroduced. The first revelation came when we climbed a lookout tower used for spotting where and how a fire may be moving through the refuge. From that bird's-eye view, I could see what I did not see anywhere when traveling from the Great Dismal to the Pocosin Lakes and beyond toward South Carolina and Georgia. There was nothing but trees, shrubs, and water as far as the eye could see. And, as I observed over the course of the next two days while visiting the Alligator River Refuge and other wetlands in the area, many white-tailed deer, racoons, rabbits, and nutria. The wolves would have to be very poor hunters to starve here.

The Pocosin Lakes, the Alligator River, and the Great Dismal Swamp Refuges were once part of a vast, semi-interconnected system of soggy peatlands that extended from New York State to Florida. The fact that one of the reintroduced Alligator River wolves made it to the Great Dismal is a testament to the fact that the peatlands are still connected by wildlife corridors, small though they may be. But these peatlands differ in some subtle and not-so-subtle ways. The trees in the Pocosins tend to be smaller and less common, and shrubby pond pine (pocosin pine) is much denser in the Pocosins than it is throughout most of the Great Dismal Swamp.

"It's the peat that determines what kind of vegetation grows here," Phillips told me. "In areas where you have peat more than four feet below the surface, you see mostly shrubs. In areas where you have peat that is four feet deep to one and a half feet deep, you get Atlantic white cedars, bays, and pond pines. Oak trees are found in areas where there is only a foot or so of peat."

Phillips paused for a moment as he did a slow 360-degree scan for signs of life.

"It may not be the best place for a wolf to make a living, but there is lots for them to eat and there is a lot of room for them to roam," he noted. "We used to have more wolves here, but we have only two left, as far as we can tell. The farms around us are the reason why they leave the refuge and get killed. It's like setting up a table of food for them out there. It's easier for wolves to make a living outside the refuge than it is this far inside."

Phillips didn't say it, but I had a hunch that he thought that this red wolf experiment might turn out just as badly as it did in the Great Smoky Mountains, where a simultaneous reintroduction of red wolves in the early 1990s ended abruptly. That time, it wasn't public opposition but an apparent shortage of food and disease that killed most of the pups.

In a way, this is personal for Phillips. He is a native North Carolinian who could easily serve as a poster boy for many residents of the state who served in the military and live to fish and hunt on holidays and weekends. But that's not how he has been treated.

On a highway outside the refuge, there is a billboard image of a bearded man dressed up in a Fish and Wildlife uniform with a red wolf hanging onto his long Pinocchio-like nose. The wolf is about to fall into what looks like a flooded forest. "*U.S. Fish and Wildlife Service, the lying neighbor no one should trust,*" the billboard reads. "*Google Red Wolf Restoration Scandal.*"

Phillips has a beard and dark hair, just like the uniformed man on the billboard. He doesn't doubt for a second that he is the target of Jett Ferebee, the wealthy landowner and real estate developer who paid for this roadside public-service-bashing advertisement. No one else in the Columbia Fish and Wildlife Service office looks anything like the image of the man on the billboard.

"It's been up for two years," Phillips told me, keen to point out how much he has invested in this job and how he doesn't deserve to be a target like this. "No one has done anything to take it down."

It is true that some people were unhappy when the US Fish and Wildlife Service chose to reintroduce wolves to the Albemarle Peninsula in 1987. They feared for their children, pets, and livestock, and for the deer they hunted. It didn't help when some wolves took to the highways, chasing cars and trucks, and even walking down the street of Manns Harbor on a day that most everyone there remembers vividly. It was the kind of naive behavior one would expect from wolves bred in captivity.

But many of the wolves began to behave as wild carnivores do, killing raccoons, marsh rabbits, muskrats, deer, and even rats and nutria, a muskrat-like rodent that was brought to the United States from South America, beginning in 1899, to foster the development of a fur industry. By the time Phillips arrived on the scene in 2001, a certain degree of pride had settled in among some residents living on the peninsula. Many of them had come together in the early 1980s, when a young man by the name of Todd Miller got a grant from Mary Reynolds Babcock Foundation to find a way of preserving North Carolina's sounds and wetlands.

Ferebee was evidently not one of them. Nor was Tom Remington, a self-described researcher and author who has refused to believe that these are real wolves. They are, he has stated in one of his many blog postings, a "a cross-bred hybrid that resulted in a semi-wild canine that is of no use to anybody or anything." Remington likens the Fish and Wildlife Service to "fascists," and blames the death of so many wolves not on hunting and highway mishaps but on the ongoing efforts to rewet the peatlands in the Pocosins.

In one case, wolf pups really did drown. But rewetting the peatlands has little to do with wolves being driven out or flooding adjacent farmlands like the one Ferebee owns. The purpose is the same as it is in the Great Dismal. The main goal is to repair the damage that was done by more than a century of cutting down trees and draining the peatlands to create farmland.

Many of the agricultural initiatives in the twentieth century were not typical family farm ventures. They were corporations such as the one owned by Malcolm McLean, owner of a number of national trucking and shipping companies. A native of North Carolina, McLean got his start in business selling his mother's chicken eggs during the Great Depression, and then buying a truck, and then more trucks to deliver goods. He was already rich when he came up with the simple but brilliant idea of using containers to ship products over land and sea. It has been described by some business analysts as the "greatest advance in packaging since the paper bag."

McLean wasn't always so successful with his investments. In the early 1970s, long before the refuge was created, he bought 400,000 acres of pocosins on the Albemarle Peninsula at a price of about $200 per acre. The plan was to grow corn, soybeans, and wheat, most of which was to feed cattle and hogs. McLean planned on producing a million head each year. The idea of being the largest hog farmer in the country somehow appealed to him, more so than the honors and accolades he got from President Ronald Regan and from *Fortune* and *American Heritage* magazines.[15]

In order to turn First Colony Farms into productive farmland, McLean's company proceeded to drain it with ditches similar to those that were excavated by enslaved people in the eighteenth century. Copious amounts of fertilizer and lime were added to supplement nutrients and reduce the natural acidity of the peaty soil. Typically, four to eight tons of lime were needed for each acre of new agricultural land. An additional ton of lime had was required each year to increase the availability of nutrients for the crops.

Fishermen noticed when the nutrients triggered algal blooms downstream. Jim Meekins of Stumpy Point, which was once a lake before a peat fire eroded its seaside banks, spoke for many of his colleagues in 1983 when he stated publicly that the "Sound is becoming a cesspool." This was once one of the biggest nursery areas for shellfish and shrimp on the Eastern Seaboard, said H. O. Golden. "It's not anymore. It's gone."[16]

Peat filters nutrients out of water. The Fish and Wildlife Service knew that, without the peat to trap the contaminants, runoff from these agricultural fields would likely threaten fish and shellfish nurseries located downstream. But that didn't persuade them to intervene, as local fisherman like Meekins had hoped. Nor did the Service do anything when McLean, who was having a tough time turning First Colony into a moneymaker, sold the land to a peat-mining venture owned by Washington, DC, insiders who had a plan to turn pocosins peat into methanol.

Investors in Peat Methanol Associates (PMA) included four former Reagan Administration officials, including CIA director William Casey and Robert Fri, who had been acting administrator of the Environmental Protection Agency in the early 1970s under President Gerald Ford. Their $576-million peat mine proposal promised to gasify 633,000 tons of pocosins peat on 15,000 acres annually in order to turn it into 51 million gallons of a gasoline fuel additive that had not yet been approved by the government.[17] The company had options on another 100,000 acres.

The venture would not likely have had a chance of moving forward had it not been for the US Synthetic Fuels Corporation (SFC), which provided $465-million in loans and price guarantees.[18] SFC was created in 1980 under the Energy Security Act in response to the 1973 oil embargo by OPEC, the Organization of Petroleum Exporting Countries. It had an initial $20-billion annual budget to incentivize production of new sources of energy. Most of the loan guarantees and guaranteed price purchases went to coal and oil companies. (Wind and solar were not on the agenda back then.) Turning coal—and peat, in this case—into synthetic fuel was the main goal.

The generous subsidies that went to Peat Methanol Associates didn't seem to be justified, given the fact that even the SFC staff had doubts about the efficacy of transforming peat into methanol. One early assessment by the US General Accounting Office described serious concerns and questions about the viability of the project and its ability to add significantly to the nation's capability to expand synthetic fuels rapidly or at a large scale in the future.[19]

But the subsidies were approved anyway, after staff analysts were instructed to "'water down' negative reports on peat [and] to put peat in a more favorable light."[20] No one knows whether SFC deputy inspector general Robert W. Gambino had anything to do with it. But he was the CIA director of security up until 1980, when Casey was Ronald Reagan's campaign manager and was about to be made director of the CIA.

Fishermen were up in arms. Tom Carroon spoke for many when he warned that ignoring what PMA was about to do was "like turning your back on a bull."[21] This turned out to be Todd Miller's big moment. A year after forming the North Carolina Coastal Federation, he rallied fishermen, environmentalists, and scientists to oppose further development of the mine.

Four conservation groups, including the North Carolina Coastal Federation, filed suit to stop the peat mine, claiming that the US Army Corps of Engineers was not enforcing federal regulations that protect wetlands. It was one of many lawsuits filed across the country that reflected a larger battle to save the nation's dwindling wetlands since passage of the Environmental Protection Act in 1973.

By that time, three-quarters of the 2.25 million acres of the peatlands that had existed in North Carolina in the early 1960s had disappeared, compelling the EPA to describe them as the most critically endangered wetlands in the country. Worried about what peat mining might do to environmentally sensitive areas, the commissioners of Dare County passed a resolution asking for the state to forbid further development.

That may not have been enough to stop it. But when a fire swept through the region during the exceptionally dry spring of 1985, there was not enough moisture left in the remaining peatlands to slow its momentum. The fire moved so fast that it killed 20 percent of the deer in the area, and injured that many more.[22] By the time it was put out, over 95,000 acres of peatland had burned, leaving behind depressions that were three feet deep and rapidly filling up with rainwater.

As that fire smoldered, Peat Methanol was already rapidly running out of both peat and money. Technical advice and equipment that the company was getting from Russians and Finns who were experts in turning peat into fuel weren't doing the trick. Officials went back to the SFC for more financial help. Their timing couldn't have been worse.

Most members of Congress had concluded by then that the synthetic fuels business was going nowhere under a corporation that was constantly under fire for conflicts of interest, political fundraising by board members, embarrassing resignations, excessive salaries, lavish travel benefits, and a poor track record for producing synthetic fuel success stories. The request from Peat Methanol was turned down, and SFC was forced to close its doors for good a year later. When it did, the *New York Times* marveled at how rare an event like that was in American politics. "Government agencies are easily born, but they never seem to die. Rarely do they even fade away. But at 5:00 p.m. today the Government's Synthetic Fuels Corporation closed its doors forever."

The fire lookout tower we climbed that day was a relic of that failed peat-mining venture. Nearby, we could see the waterlogged ditches that were dug by the company before the enterprise collapsed. It could have been a lot worse, said Phillips. "Most everyone agrees that there would have been almost nothing left of the pocosins by 1990 had the peat mine gone ahead. That's when the land was finally bought by the Conservation Fund and the Richard King Mellon Foundation, before turning it over to us (the Fish and Wildlife Service) to create a national wildlife refuge."

As Phillips was talking, the wind kicked up, sending in a putrid smell that clearly did not have origins in a wilderness like this one. I was afraid to ask where it was coming from because Phillips kept on talking as if nothing was untoward. When we got back to the truck, Phillips unwrapped one of his sandwiches and started to eat. I didn't join him because the smell was by this point nauseating.

A few miles down the road, the mystery that had been perplexing me was solved. "That's Rose Acre Farms," Phillips said, pointing to twelve high-rise confinement buildings about a half mile away. "It's the biggest egg production facility in the state. Up to four million hens. They lose about 1 percent a week. That's about 4,000 birds.

"There's an interesting story behind this that you should look into," he said before turning away and walking toward a water-retaining structure along the canal we had parked by to see if water needed to be held back or released.

I was intrigued. The story, I learned later, is that Rose Acre Farms bills itself as "family-owned, with small-town values." Its website quotes from the book of Psalms as if to put an exclamation point behind that. The company got its start, as Malcolm McLean did, as a family farm selling eggs. By 1939, the Rust family had three henhouse operations in Indiana that could hold as many as 900 chickens. It quickly expanded to become one of the largest egg-producing operations in the United States.

Owner David Rust was 58 years old in 1984 when he told a business journal that he wanted "more children"—he already had seven—"and more chickens."[23] His wife was biologically unable to oblige him on that first wish. So Rust left her for a Polish exchange student before embarking on an egg farm expansion spending spree that rivaled almost anything seen up until that time. He vowed to build a new plant every year for the next decade.

Rust was an eccentric. He built a seventy-foot-high treehouse so he could get a panoramic view of the Indiana operation. He ordered staff to put up American flags all over his properties and print Bible quotations on the egg cartons. He was generous with some employees, but brutal when it came to dealing with those who were caught smoking or found to be a minute late for work. The most insignificant things riled him no end. In 1981, pullet manager Ken Cone had had enough of his belligerent eccentricity. He sued for damages when Cone cut his pay for allowing weeds to grow behind his house, and for not wearing his chicken feather pin when taking his family to the local county fair.

According to employees, Rust once bit the beak off a live chick to underline the importance of trimming beaks to prevent chicks from pecking each other in such crowded conditions. He once set a table with chicken manure on the plates to prove that the manure didn't smell if it was sprayed with deodorant.[24] He was so aggressive in his expansion efforts that competitors in California launched a multi-million-dollar advertising campaign to try and slow him down. The government, he told a congressional subcommittee at a hearing that preceded the Egg Industry Adjustment Act of 1972, had no business controlling the industry because people were starving and "because if we are to continue our Christian leadership in food production in a starving world, we must remain free."[25]

Rust's long-suffering wife and seven children got their revenge in divorce proceedings that allowed them to take over the company, making it even bigger and more efficient than ever in expanding its market share. The company's eighteen egg farm operations in the United States are often in court fighting charges of price-fixing, labor relations infractions, and environmental violations, or dealing with the Food and Drug Administration over health and safety issues.

The facility on the edge of the Pocosin Lakes National Wildlife Refuge is no exception. In 2018, the FDA cited the North Carolina operation for unacceptable rodent activity, unsanitary conditions, and poor employee practices that create an environment that allows for the "proliferation and spread of filth and pathogens throughout the facility."[26] In a Q&A in 2008, Marcus Rust, the executive vice president of the company, said they chose to build in North Carolina because of proximity to markets and a long distance from other egg operations, and because the peaty soil "was poor in nitrogen, suggesting that disposal of manure would not be a problem."[27]

There's nothing in the Bible that tells you what to do with urine and fecal matter produced by three to four million hens. The Rose Acre Farms operation uses ventilation fans to blow out about 2.3 million tons of ammonia, nitrogen, and phosphorous annually. Much of that gets absorbed into the pocosins and the Tar–Pimlico river system. The Pimlico–Tar River Foundation, Friends of Pocosin Lakes National Wildlife Refuge, and Waterkeeper Alliance took the company to court several years ago.[28] The fans, however, were still blowing when I was there.

When Howard Phillips isn't defending the red wolf program, he's doing his best to explain to people why it's so important to keep peat in the pocosins wet. That should have been an easy job after several peat fires burned for months on end in 1985, 2008, and 2011. The 2008 Evans Road fire was especially troublesome. It required almost $20 million in fire suppression efforts. Smoke from those fires triggered a spike in the hospitalization of people suffering respiratory problems. It also limited the amount of training that could be done by navy and air force pilots over the nearby bombing range.

Building water-retention structures, as Fred Wurster and his associates were doing in the Great Dismal Swamp, is the most cost-effective way of keeping that peat from igniting when lightning storms pass through. It's not their primary purpose, but the weirs also hold back rainwater that has the potential to flood farms and towns that surround the refuge. And then there is the tantalizing potential that money can be made on carbon offsets by keeping the carbon in the ground.

"I'm a landscape plumber, not a hydrologist like Fred Wurster," Phillips said as he lifted one of the slats on the water retention structure to allow more water to drain out. "Water on these canals can go both ways, depending on weather and how we control the flow. It's intuition to me as much as it is a science."

Once again, he stopped in mid-thought and looked around.

"I'm sensing that we're drier here than we should be. I wish I could get Fred Wurster to come down here more often so I could pick his brain. But we just don't have a budget for that these days. We have half the number of staff there were when I started working here."

Rising sea levels and an increase in tropical storm activity is making it hard for Phillips to convince skeptics who have suffered serious losses from the flooding that has come with a dramatic uptick in hurricane and tropical storm activity along the southeast coast of the United States. Even though only 18 percent of the refuge is being rewetted, they blame him and the Fish and Wildlife Service when floods cause them grief.

"In 2016, Hurricane Matthew dropped a record eighteen inches of rain on us in just 36 hours," he said. "That's almost as much as we normally get in our wettest month. Twenty-five people died. More than 100,000 homes and business were damaged. There was almost $5 billion in damages. It would have been a lot worse had we not done all we could to hold back as much water as we did. Instead of thanking us, they blamed us. It wasn't just the Pocosins that got flooded, I tried to tell them. It was the entire state and most of the Southeast Coast."

Walking down to the banks of the canal to better see how Phillips was controlling the flow of water, I saw out of the corner of my eye something long and dark slithering into the tall reeds. It had been lying there no more than two feet from me. Seeing how I froze in my tracks, Phillips asked me whether I heard a rattle. I didn't and probably wouldn't have heard if it did rattle, because the wind was really humming by this point. "Could have been a mud snake," he said. "We got a lot of snakes here, but it's a little early for them to be out at this time of year. In another few weeks, they will be a lot more lively."

The list of animals in the Pocosin Lakes is a long one that includes 20 species of snakes, 16 species of salamander, 21 species of frogs, and 8 species of turtles, as well as alligators, river otters, deer, fox, bobcats, and the densest population of black bears anywhere in the world. Winters are so warm here, thanks to the Gulf Stream that bounces off the peninsula, that the bears have no need to hibernate.

As we continued to drive along, the wetland scenes we saw resembled one of those idyllic television nature shows—the dark shadows of bears lurking in the forest, deer bouncing through the shrublands, muskrat swimming, and beaver tails slapping the surface of the water. I laughed out loud when I saw a turtle and nutria sitting side by side on a tiny peat mound in the middle of a bog. The Disney Company could have made a story out of that.

It's the bird population that really stands out here. In the pocosins and other peatlands in the Albemarle, there is everything from the little chuck-will's-widow, which has a mouth big enough to swallow whole a songbird half its size, and black vultures, which have been known to kill newborn calves and lambs. There are owls, hawks, eagles, and peregrine falcons. As many as 100,000 snow geese and tundra swans spend the winter in the region's 7,300 acres of lakes, ponds, and impoundments.

The one thing that struck me as odd is that the Fish and Wildlife Service pay farmers to leave part of their crops unharvested for snow geese, tundra swans, and other waterfowl to consume. The last thing the peatlands of the Canadian Arctic need are more snow geese, which are destroying large swaths of tundra ecosystems. The snow goose populations have exploded because so much of the wetlands in the American Southeast have been drained and converted into cropland.

Driving along, Phillips suddenly ducked to get a better view of some geese that were flying overhead. "Did you see that?" he asked.

"I did," I answered, "but was reluctant to admit that what I saw were Canada geese, a bird that is so common where I live, it is considered to be a nuisance."

"Really," he said incredulously when I told him that. "They are so rare here that hunters consider them a prize. I'd love to shoot one."

When we got back to the Fish and Wildlife Service office in Columbia toward the end of the day, Alligator River refuge staff and wolf-recovery biologists were meeting with an official from the United Nations who specializes in getting reluctant communities to buy into endangered-species recovery programs. The plight of the red wolf was on the agenda.

After brief introductions, Phillips walked me to my car and allowed that he had enjoyed the day. "I love this job and what it aims to do," he told me. "But I'm going to retire at the end of the year. We just don't have the resources to do what needs to be done. And I'm tired of getting beaten up by people like Ferebee. If they want to bring me back on contract for short-term projects, I'll give it some serious thought. We'll see."

In the months that followed, I learned that Phillips did retire and that his story was a common one. Most of the government-managed peatlands in North America are run by individuals who do not have the resources and expertise to do what is needed to restore bogs, fens, and swamps that have been badly degraded. The ecohydrologists and peatland ecologists who are needed are a rare breed. Few, if any, are employed by the national parks services in the United States and Canada. The states and provinces are generally uninterested, leaving restoration and management to local volunteer groups and nongovernmental organizations such as the Nature Conservancy.

Chapter 4

Tropical Peat

To think that plants ate insects would go against the order of nature as willed by God.
— Swedish naturalist Carl Linnaeus (1707–1778)

We were in the heart of the Alaka'i Swamp on the Pacific island of Kauai being serenaded by the trill of the crimson-colored Apapane honeycreeper when a skull came into view. It was strange to see it sitting there on a rock in the middle of the Kawaikoi, a boggy stream the color of black tea. It seemed to be staring at us.

I suspect my imagination was running rampant in this tropical surrounding. The forest was still dripping with rain that had fallen earlier in the day. Was the skull the head of a rat, a mongoose, or something else? On closer inspection, it looked like it might have been the head of a cat, one of an estimated 5,000 feral felines that are, along with untold numbers of rats, feral pigs, goats, and deer that don't belong in this verdant wilderness, the biggest threats to Hawaii's 424 federally threatened and endangered plant species.

Mongooses were introduced to Hawaii in a disastrous attempt to control the spread of rats in cane fields. They haven't established themselves on Kauai, the oldest of the islands. At least not yet. But many have been trapped and euthanized because of their voracious appetite for native birds. This cat, if that is what it was, may have drowned trying to hopscotch across this boulder-strewn stream when it was swollen with rainwater, in pursuit perhaps of one of the honeycreepers we were listening to that day.

Tropical peatlands are typically associated with South America, Africa, and Southeast Asia, where they represent about 10 percent of the world's peat but store between 10 and 30 percent of the world's carbon.[1] Early accounts of Southeast Asia's tropical peat swamp forests dismissed them as ecologically impoverished. More recent studies suggest quite the opposite. There are at least 1,524 plant species, 123 mammal species, 268 birds species, 219 freshwater fish species, and untold numbers of species of invertebrates, bryophytes, ferns, and fungi in the peatlands of Southeast Asia.

Many of the world's critically endangered species and several new ones have been discovered in tropical peatlands in recent decades, including an exceptionally rare orchid found here in Alaka'i, for example; an amoeba, unlike anything seen before, in the Amazonian peatlands of Aucayacu;[2] and several undescribed blackwater fishes in the North Selangor peat swamp forest in Peninsular Malaysia.[3]

We know so little about these tropical places that it wasn't until 1999 that scientists learned that fruit-eating Sumatran orangutans occasionally feast on slow lorises, often referred to as the world's cutest animal even though it is the world's only venomous primate.[4] (Lady Gaga had planned on featuring one in a music video before it bit her.) The black-browed babbler, a bird that hadn't been seen in 170 years, was found in the rain forest of Indonesia's South Kalimantan province in October 2020.

The fact that one of the world's largest tropical peatlands, in a remote part of the Democratic Republic of the Congo, wasn't discovered until 2017 further underscores just how much more there is to learn about these ecosystems and how much is being lost to invasive species, deforestation, and wildfire. Prior to 2010, when the conversion of peatlands into farms and plantations was accelerating, more than third of the peatlands in Southeast Asia had already been lost.[5]

Alaka'i stands out because it is home to more critically endangered species (and a handful of recently discovered new ones) than almost any other ecosystem of its size. Native Hawaiians refer to Alaka'i as

Kawaikoi stream flows through the boggy interior of the Alaka'i Swamp, home of some of the world's rarest plants, including the rarest orchid. (Photo: Edward Struzik.)

"Swamp in the Sky." Alaka'i, however, is not a swamp, but a tropical montane forest brimming with twenty or more bogs. It is nine miles long, two to three miles wide, and approximately 4,000 feet above sea level. It plateaus just below Wai'ale'ale, a deliriously dismal shield volcano—dismal, that is, only because it is almost always cloaked in moisture-laden clouds that form when cool trade winds that have traveled over 2,000 miles of ocean are forced upward when they collide with the warm coastal cliffsides. As the winds funnel up and cool even more, the moisture turns to fog or rain clouds that are eventually wrung out.

On average, 466 inches fall on Wai'ale'ale annually. That is almost as much as falls in Mawsynram and Cherrapunji in India, the rainfall capitals of the world. Most of the rain there comes with seasonal monsoons. The rain clouds over Wai'ale'ale are more like supermarket weather machines that, with some frightfully violent exceptions, spray a gentle stream of precipitation most days of the year. It comes as no surprise, then, that Wai'ale'ale literally means "overflowing water."

One of its steepest slopes—the Weeping Wall—ripples with ribbons of cascading streams.

Alaka'i and other tropical peatlands in Hawaii—the Pepe'opae Bog on Moloka'i, Big Bog and Flat Top Bog on Maui, and the Kanaele Bog in the lowlands of South Kauai—are among the most fragile and vulnerable ecosystems in the world.[6] Many of the other island bogs, such as the cluster found on the lower east slope of Mauna Kea, have only recently been mapped. New identifications such as *Microspora pachyderma*, one of a number of green algae found in Alaka'i, continue to be made in these places.[7]

Peat is eight feet thick in some parts of Alaka'i. There is, however, no sphagnum moss as there is in some of the other bogs in Hawaii. Nor are there many orchids, as I had expected. Just three native species. *Platanthera holochila,* a twenty-inch-tall orchid whose ancestors likely originated in the cold bogs of Alaska, is the rarest.[8] Botanist Steve Perlman found it on a typically foggy day in Alaka'i.

What there are a lot of are weird, exotic, and improbable plants, such as *Cyanea kuhihewa*, a rare Hawaiian member of the bellflower family, that Ken Wood, Perlman's field partner, found more than thirty years ago. It is a spiky green plant with purple, tongue-like flowers bursting from the trunk like fireworks. It was thought to be extinct until Wood found one in 2017 in a remote forest on the island.

My wife, Julia, and I have embarked on many challenging adventures over the years. Among them was the first descent of the Nanook River on Victoria Island in the High Arctic, a month-long journey across Banks Island on the most northerly navigable river in North America, and a paddle down Canada's South Nahanni River, which is to some paddlers what Everest is to mountaineers.

Hawaii has never been high on our list of places to explore because it is so crowded with tourists. Julia, however, was intrigued when I suggested the idea of hiking into Alaka'i, selling her on the fact that it is a refuge for hundreds of plant, insect, and bird species that are found nowhere else outside the Hawaiian islands. It's a little like the Galapagos, a place where evolution has run rampant, allowing plants and birds that migrated from the mainland hundreds of thousands and even millions of years ago to develop new lifestyles and evolve into separate species.

Initially, we were enticed by the prospect of seeing a 7.5-inch-long dragonfly with a wing span of four inches (pinao), a carnivorous caterpillar that snatches flies out of the air (*Eupithecia*), the very rare "fabulous green sphinx moth," Lehua blossoms on the native ōhi'a trees, the island's endangered short-eared owl (pueo), and a critically endangered honeycreeper (akikiki) that is rarely found outside Alaka'i.[9]

But we began to have second thoughts as we sorted through historical accounts and reports of more recent trips into the region. In 1865, botanist William Brigham couldn't find anyone who would take him there because it abounded in "deep mud holes, and several natives had lost their lives there."

"Oozie, eerie," and "crazy muddy trails" were among the most common descriptions that modern-day hikers used to describe their experiences. "Bushwhackers lost for days or plumb vanished," warned another. "The swamp made us feel like we were in the Dead Marshes of *Lord of the Rings*," was one that stood out.

Whatever doubts we had were dispelled by Kathryn Hulme, the author of the 1956 best seller *The Nun's Story*. Hulme lived her life out on the island. Alaka'i, she wrote poetically in the *Atlantic Monthly*, was "an antediluvian world of quaking bogs and stunted as well as giantlike vegetation, where violets turn into trees, trees into ground shrubs, and every sense you ever hear about customary nature is turned upside down."[10]

I had ambitious plans for us to hike through Alaka'i to Wai'ale'ale a week before I was to begin a week-long immersive program at the National Tropical Botanical Garden with fellow writers from *National Geographic*, *Smithsonian*, and National Public Radio. But hopes of reaching the summit to pay homage to a shrine dedicated to Kāne, the Hawaiian god of living creatures, were dashed when we learned that landslides, and trees downed by a weather bomb that dropped fifty inches of rain in 24 hours the year before, made the hike virtually impossible.

It was hot, foggy, and raining intermittently when we set out along the Pihea Trail, one of the more traditional starting points on the northwest side of the island. Studded hiking shoes gave us a reassuring grip on slippery slopes of red clay. Narrow boardwalks got us through some of the spongiest parts of the Swamp. A wooden staircase that took us down a steep slope was a godsend, rickety and missing steps though it was.

The hike was not the slog it would have been had heavy rain flooded the area. What made it such a pleasure for boreals like us was the absence of the mosquitoes and biting flies that can make a trek through most peatlands a maddening experience.

As we moved in and out of fog through ferns (hapuʻu and uluhe lau nui), flowering evergreen trees (ōhiʻa), tropical sandalwood (iliahi), spongey mosses, and rotting wood, I was reminded of the film *Jurassic Park*, which was shot in a botanical garden on the south side of Kauai.

There are no monsters thundering through the rain forest in Alakaʻi, as there were in the fictional Jurassic Park, just enormous feral pigs that lurk along the perimeters, black deer that were introduced to the island in 1961, and goats that arrived with European visitors such as James Cook and George Vancouver. There are reportedly more pigs on Kauai than people. The backside of a boar was the only one we saw. It immediately gave us the right of way, disappearing so silently into thick shrubs that I thought I had imagined seeing it.

What we certainly did see were luminescent, lime-green crab spiders, predatory lacewings that act like ground beetles, crickets that trill and apparently chirp at night, and many small birds flitting from ōhiʻa tops to lapalapa, six- to fifteen-foot-tall trees of the ginseng family that quiver like aspen do in the boreal forests. We puzzled over the white egg-shaped structure of sea anemone stinkhorn, a fungus that secretes a vicious spore-laden gel (gleba) that smells like rotting meat or worse when it fruits.

The human history of Kauai began with the Polynesians arriving by canoe 1,200–1,600 years ago with nothing to guide them but the stars and the trade winds. When Captain James Cook landed in Waimea, on the west shore of Kauai, in January 1778, European missionaries and settlers quickly followed.

The first recorded ascent into the Alakaʻi was made in 1821 by two missionaries—Hiram Bingham and Samuel Whitney, long before

William Brigham tried and failed to get someone to take him up there. Whitney was "charmed by the melodious singing of birds' notes; I am sure sweeter than the strains of Orpheus."[11] The men were so often diverted by the scenery that their Hawaiian guides grew impatient. Their enthusiasm waned, however, when the footing became so bad they had to get down on their hands and knees to get through the mucky peat.

American botanists Charles Pickering and William Brackenridge found the going just as tough when they arrived on the island in the fall of 1840 with the United States Exploring Expedition led by Lieutenant Charles Wilkes. They described the landscape as being "filled with quagmires" at the bottom of slopes so "very steep and fatiguing; but by slipping, tumbling, scrambling, and swinging from tree to tree" they found their way out.

Valdemar Knudsen, an engineer of Scottish and Norwegian ancestry, was one of more beloved of the early arrivals, possibly because he eagerly mastered the local language. He settled in Kauai with a plan to raise cattle on a failed tobacco plantation that he bought sometime in the early 1850s. Knudsen's real passion was to go for long hikes into the deepest, darkest places with his guide, a man named Kaluahi.

It was Kaluahi who guided Hawaii's Queen Emma through the Alaka'i Swamp on horseback following the death of her only son, Crown Prince Albert, in 1871. The queen evidently found the cure she was hoping to find for her sorrow when her entourage reached a high point just above the Swamp. Enchanted by the vista, she ordered the dancers and chanters she had brought along to perform the traditional Pohakuhula, what most of the world now knows as the hula dance.

When Queen Emma triumphantly returned from her trip through the Swamp to the north shore of the island, songs and chants were written to commemorate her journey. Protocols were put in place to deter people from picking the leaf buds of the ōhi'a tree. Throwing stones at small birds was deemed a sacrilege.[12] The Alaka'i Swamp became a sacred place.

Stories of little people added to the mystique of the place. Like the Pulayaqat of the tundra in the western Arctic, and the pixies of the moors of Britain, the Menehune of Hawaii possessed extraordinary powers. Legend has it that even as they walked in relative silence, their murmuring voices could be heard across the ocean to Oahu, the island to the west.[13] On the instruction of Ola, a ruling chief of the island, the Menehune reportedly built Kipapa-a-Ola, a corduroy path of hapu'u logs through the nastiest regions of the Alaka'i Swamp. Seeds from the hapu'u sprouted, lining the trail with lacy fronds of ferns.[14] The Menehune, however, refused when some other powerful person asked them to divert water from a wetlands.

What makes Alaka'i critically vital is that it is, along with Wai'ale'ale and the steep and inaccessible cliffsides of the northwest coast, the last best hope for many of the native species in Hawaii that are on a fast track to extinction. Among the 1,400 vascular plants that are native to the Pacific state, 90 percent are found nowhere else. More than a hundred are extinct. Four hundred and eighty-one are listed as endangered.

The story of rapid extinction in Hawaii began with the Polynesians arriving with rats, pigs, and root vegetables, followed by Captains Cook and George Vancouver who brought goats, mosquitoes, and even-more-aggressive rats. When the European missionaries and settlers arrived, they came with rabbits, cats, cows, and birds as well as axes, spades, plows, plants, seeds, and biological controls to produce crops such as tobacco and sugarcane.

Spreading seeds and plants around the world had become a common and often profitable practice since the days when John Bartram and his son William were harvesting them in the peatlands from the Pine Barrens of New Jersey to the Everglades to send to collectors overseas. An estimated 13,000 plants have been introduced to the Hawaiian islands. Many native species couldn't compete with the new arrivals because they had lost pre-existing defenses such as toxins and thorns to ward off the intruders. There was no longer any need for these survival tools when wind and water brought Hawaii's original flora to these benign shores. They used that energy instead to adapt to novel ecological niches.

Ground-nesting birds such as the Laysan finch didn't know how to respond when rats, cats, dogs, and mongooses began to prey on them and their eggs. Tree-nesting birds were just as helpless fending off the arboreal-foraging habits of the black rat, which can climb trees and reach densities of 119 per hectare.[15] Yellow-faced bees became easy prey for alien ants and nonnative predatory wasps. Plants that required the services of pollinating bees and moths stopped producing seeds. Several native Hawaiian honeycreepers were driven to extinction by mosquitoes that transmit avian malaria.

When the songbirds began disappearing, organizations such as Hui Manu, a ladies' club whose sole purpose was to introduce songbirds to Hawaii, brought in northern mockingbirds, Japanese white-eyes, and Japanese bush warblers. The "buy-a-bird" campaign that began in 1928 encouraged schoolchildren to raise money for the introduction of the northern cardinal, the hwamei, and the red-billed leiothrix. These and the varied tits, indigo buntings, and blue-and-white flycatchers that followed drove even more native species out of the lowlands into the mountains and boglands of Alaka'i.

As bad as it turned out to be, it would have been a lot worse had many of the introduced birds successfully established breeding populations.

The US Geological Survey investigated the possibility of harvesting peat in Alaka'i for fuel, only to give up when officials realized that Alaka'i was too far away from markets. The federal government had plans to build a hydroelectric dam in Waimea Canyon that would have flooded most of the Swamp. The idea was shelved when Stewart Udall, the US secretary of the interior from 1961 to 1969, was brought in to listen to public concerns. Udall reportedly needed no more convincing when he was taken to the Kalalau Lookout at Waimea. "Nowhere in the entire national park system is there scenic beauty such as this," he reportedly said. "Give me the proclamation [to turn it into a national park], I'll sign it now."[16]

In recent years, the integrity of Alaka'i has been challenged by forces from within the state. To the dismay of conservationists, the island's hunters have opposed efforts to eradicate the goats that are chewing up rare plants and their habitat at an alarming rate. (One study indicated that 98 percent of the goats' diet consists of native plants.) Once the culling was initiated, hunters made their feelings known at public meetings, on television, and in newspapers. The state finally gave up when a group of them ventured into the Kalalau Valley with machetes, defacing three of the rarest trees on the island so badly that they eventually perished. The word *die* was carved into the trunk of one of them.

Many native species maintain a foothold in Alaka'i because it is too high and cold for mosquitoes that infect birds with avian malaria, and the terrain is too wet and muddy to hold the weight of roads that divert or block the flow of water. And like most boggy places, people don't come to the Alaka'i Swamp in droves.

Alaka'i is not just a refuge for birds and plants, it's a fortress that has been designed to keep the enemy out. Fences have been built to prevent feral pigs, exotic deer, and goats from debarking trees, eating plants, or digging up, wallowing in, or trampling the spongy areas where many rare plants grow. Inside the fence, the rarest of plants have enclosures around them to give them added protection. Even with all that, invasive plants such as the highly aggressive roseleaf raspberry are finding their way in.

Still, everywhere we looked there was something new for us to see and marvel at—red-berry-filled pukiawe bush (*Styphelia tameiameiae*), 'a'ali'i, a native soapberry shrub, and kanawao, which is a member of the hydrangea family.

I was tying my shoe when a more familiar plant caught my eye. *Drosera anglica*, the carnivorous sundew, is typically a cold-climate species that is dormant for a good part of the year. What it was doing here in a tropical montane rain forest was a mystery that was later solved when I got a chance to go for a hike with Steve Perlman and Ken Wood—two men who have done more for the discovery and preservation of rare and endangered plants in Hawaii than anyone else.

"Extreme botany" is a term botanist Steve Perlman uses to describe what he and colleague Ken Wood do to save the rare plants that grow in Hawaiian bogs, along steep cliffsides, and in the deepest, wettest parts of the tropical montane forests. If they aren't bushwhacking through bogs and thick brush, they're climbing up or rappelling down those vertical walls. In many cases, the only way of getting to their destination is by ocean kayak or helicopter.

"It's always an adventure, and not one for fainthearted," Perlman told me as we were driving from the National Tropical Botanical Garden in the south to Pihea trailhead in the mountainous northeast, where there are several starting points for a hike into Alaka'i.

Perlman chuckled when I asked him how it felt being called a "rock star," as some have done.

"It's not how I see myself," he said simply.

Perlman and Wood are to Hawaii and the islands of the South Pacific what the Bartram father-and-son team were in their day to the cedar swamps on the Atlantic coast. Like the Bartrams, their collection of rare and undiscovered plants is unparalleled in scope and importance. Without the Bartrams, many of the most vulnerable species would have disappeared without notice. Both went to extremes joyfully, as few could or would have done in their time.

Steve Perlman was 71 years old when I met him. He was stout and fitter than most people half his age. He had the weathered look of a surfer who has been in the sun too long. Surfing, I learned on the drive, was still a passion for him. "It's gotten harder and harder to find a place where you can surf alone or with just a few people around," he said as we passed a row of cars on the highway where people have carved a trail to a remote part of the coast with good surf and few tourists. "Now, more than in the past, you have to choose your spot and go at a time of day when it's not so crowded."

Perlman spent 42 years working for the National Tropical Botanical Garden before joining the Plant Extinction Prevention Program. PEPP is a small organization that it is trying to save 238 native Hawaiian plants that have fewer than 50 specimens remaining in the wild. Eighty-two of them are on this island.

"It can be depressing work when you consider that a loss of seven or eight plants can result in extinction," he said. "When they do finally go, it's like losing a member of the family. But it can also be wonderful when you hike into places like Alaka'i or Wai'ale'ale and find species that are either new or thought to be extinct."

One of those rare finds was a palm that Perlman spotted on slopes of Wai'ale'ale. *Pritchardia perlmanii*, the plant that now bears his name, is 12–15 feet tall. There are just 200 left in the wild, a number Perlman considers to be "plenty" by Hawaiian rare-plant standards. An even rarer find was *Platanthera holochila*, the rare orchid he found in the Alaka'i Swamp in 1977. The pregnant seeds that he gathered that day were eventually propagated with great difficulty before be-

Botanist Steve Perlman describes what he does to save critically endangered plants in the Alaka'i Swamp and other remote areas of Hawaii as "extreme biology." (Photo: Edward Struzik.)

ing planted in a fenced enclosure in the swamp out of the reach of predators.

Perlman's career got its start in 1971, when he was landscaper at the Olu Pua Gardens in Kauai. When the owner expressed a desire to develop a native plant collection, Perlman saw opportunity, so he applied for and got an internship at the National Tropical Botanical Garden. It was there that he met his mentor, Derrel Herbst, a botanist who had a keen interest in rare native plants.

"Derrel had a fear of heights and didn't like to climb trees or scale cliffs, so I would do it for him," he recalled. "It was Derrel who taught me the art of collecting seeds."

Perlman's first of many life-changing moments occurred during a solo hike in the mountains, where he found five extremely rare plants. The thrill of that discovery convinced him to go back to school to pursue a graduate degree in botany.

When Perlman returned to Kauai, he met up with Harold St. John, a former University of Hawaii botanist whose many claims to fame included a successful search for an alternative source of cinchona in the Andean rain forest. The bark of the tree contains quinine, an anti-malaria drug that was in short supply during the Second World War.

St. John and some of his students explored the Alaka'i Swamp in the summer of 1930, identifying and eventually writing monographs on a number of endemic species. St. John was convinced that there were new discoveries to be found in the bogs and on the steepest cliffsides he could not get to. When he learned that Perlman was a rock climber, he encouraged him to give it a try.

Perlman took up the challenge, using little more than a 10-millimeter rope and a hard hat to protect him from falling stones. "The rarest plants tend to live in areas where pigs and goats can't get to them," he told me. "Those are the places in the world that are our best hope."

Perlman has found many rare and undiscovered plants this way. Finding alulu (or ōlulu), the so-called cabbage on a stick (*Brighamia insignus*), was one of the highlights of those early cliffside adventures.

"They were flowering when I found them, but they weren't setting much fruit, likely because their pollinator, the fabulous green sphinx moth, we think, was almost extinct. So we started to pollinate them.

We're down to just one in the wild, but we have thousands of them growing in botanical gardens around the world that were planted with seeds we sent them."

By his own estimation, Perlman has discovered approximately fifty new species and rediscovered some that were thought to be extinct. His name is affixed to ten of them. It's not just the plants that interest him. "On their own, they're just museum pieces," he says. "It's their network of relationships that must be preserved. You can really see how that plays out in places like Alaka'i, where there is a lot of wilderness that has not been disturbed in any significant way. That's the future."

Perlman smiled again when I told him about finding *Drosera anglica* in Alaka'i. He's seen it many times. Kauaians, he informed me, have a name for it—*mikinalo*—which means "to suck flies."

"Likely got there with seeds attached to a migrating bird," he said. "We have golden plovers that spend the summer in Alaska, where that sundew grows. The birds winter here in Hawaii. It's a long shot that seeds from a plant brought in by birds or wind will successfully grow here. But it does happen; some estimate it's every 10,000 years on the islands."

Along Waimea Canyon road, which climbs up the northwest coast, Perlman stopped the van to describe the many ways in which this network of relationships in the Hawaiian ecosystem has been upended since the days when Polynesians arrived with pigs and medicinal plants as well as fruits and vegetables, and when Europeans followed with cattle, sugarcane, exotic timbers, and other nonnative plants.

"The landscape changes when farmers settle in a forest," he said, reminding me of what I had already seen in the Great Dismal Swamp and the Pocosins of North Carolina.

Looking back down on the Waimea Valley, where marshlands were drained more than 150 years ago to grow sugarcane, and where genetically modified corn now grows in some spots, Perlman noted with some despair that not a single native plant could be found in our wide field of view.

As we drove farther up the mountain road, the number of native plants we saw increased. But with each year that passes, invasive species have climbed another foot or two. Part of it is the aggressive nature of these plants. Climate change spurs them on with rising temperatures that favor them over native species.

"I used to see a lot of white from the hibiscus that once grew over there," he said as we stopped again to look across Waimea Canyon. "Now it's mostly silk oak, which was spread by seed planes in the 1930s because foresters at the time felt that the native forest didn't look healthy. It's just another of the many misguided attempts by humans to introduce more species to Hawaii. Every story like this has ended the same way: native species, which have evolved over hundreds of thousands of years, get choked out by invasive plants or the get eaten by pigs and goats."

There's a fear that Alaka'i will someday suffer the same fate as Kanaele Bog on the south side of Kauai. The eighty-acre wetland was literally overrun with feral pigs, rats, cats, and untold numbers of invasive plants. In an effort to bring the lowland bog back to life, The Nature Conservancy got permission to fence it in. Volunteers pulled out 90,618 weeds to help the bog recover.

More than 82,000 of those plants were strawberry guava, a Brazilian plant that was introduced to Hawaii in 1825 for its fruit and ornamental features. Without disease or predators to check its growth, it spread quickly. Dense thickets of guava crowded out native plants that offered food for birds and insects.

"Pigs and birds love it," said Perlman. "It hasn't got a foothold in Alaka'i yet, but kahili ginger has."

Kahili ginger is in the top 100 of the International Union for Conservation of Nature's list of worst invasive species. Another is Guinea grass. It was introduced to the island 25 years ago as a means of feeding cattle. The cattle industry, however, never materialized in any meaningful way. The so-called sourgrass, which can grow to heights of eight feet, has now taken over many parts of the island.

"When and if the Guinea grass burns—and it will, because we have kids here who love to light fires—it could take out the last of a native sunflower [*Wilkesia hobdyi*] that grows on the side of a ridge just below us," said Perlman. "*Melanthera waimeaensis* might survive a fire, because it grows on the steep slopes here. But there are only seven of them left."

As we headed back to the van, Ken Wood pulled up with other members of our group. Perlman grabbed both of Wood's hands and greeted him warmly as an old friend he hadn't seen for a while.

Wood was wearing a white, long-sleeve T-shirt, camouflage pants, and a hat that he had on backwards. His gray goatee was the only feature that suggested he was 65 years old.

There's a Zen-like quality to Wood's disposition that suggests a certain oneness with this wild world that he and Perlman have explored. Walking with him as we visited a number of fenced and locked enclosures farther up into the highlands, he verified my impression when I asked him what it was that compelled him to continue his work when the odds of success are so stacked against so many of these native species.

"It's not just about me or my physical being that motivates me," he said. "It's what I am a part of. I think that deep down we all strive to understand our relationship with the Earth and the universe. The more I learn about that story, the more I want to turn the pages to see how this incredible story plays out. Being part of that story is pretty cool. To me, it's self-evident that all life-forms should be treated equally and be granted habitats like this to increase and to not be disturbed."

Wood was still flying high after he and Ben Nyberg, a drone specialist working for the National Tropical Botanical Garden, had found a plant earlier in the year that was thought to be extinct. Hua kuahiwi (*Hibiscadelphus woodii*), a relative of hibiscus, was growing on a cliffside in Kalalau Valley, just a few miles west of Alaka'i.

Hibiscadelphus woodii, or Wood's hau kuahiwi (Hawaiian for "snow mountain") was named after Wood when he discovered it in the same valley in 1991. It is a shrub that produces bright yellow flowers that turn purple as they age. No one knows for sure, but they are likely pollinated by native honeycreepers such as the 'amakihi, which is faring well on the island.

When we arrived at the Pihea trailhead, where Julia and I had started our hike into the Alaka'i ten days earlier, the wind was blowing in from the ocean. The fog and light rain was like a sheer curtain blowing in the wind. Several hikers we met had abandoned plans to venture into the Swamp.

As we stood on the hilltop, looking down into the Alaka'i, Perlman recalled the day he found that rare orchid.

"It was wet and foggy. The chance of finding a twenty-inch-tall plant that had not been seen in sixty years was highly unlikely, especially in waist-high shrubbery. When the seemingly impossible presented itself, it was ripe for picking up pregnant pods filled with seeds."

Perlman sent the seed to a network of orchids nurseries and tissue culture specialists from all over the world, hoping that one of them might be able to propagate it. But everyone who tried failed.

The promise of success came in 2000 when Lawrence Zettler, an Illinois College researcher who specializes in using mycorrhizal fungi to grow rare orchids, took on the challenge. It wasn't fungi that did the trick but asymbiotic germination, which, at first, Zettler didn't think was up to the challenge.

Perlman vividly remembers the day in March 2011 when he, Zettler, and others hiked into Alaka'i with eight plants. "It was a clear day, the first in several weeks where it hadn't rained heavily. We hiked in and planted it in a small enclosure. Thirty or more years had passed since I first found it. It was one of those rare moments when you realize that it's worth trying even when success doesn't seem like a possibility."

On the drive back, I asked Perlman what was next for him.

"I'm going to retire soon," he said. "We just got notice that the budget of the Plant Extinction Prevention Program is going to be cut yet again by the federal government. I get to keep my job, but it requires me going to the Fish and Wildlife Service office every day. A big office building is no place for me. I'd prefer to hand this over to someone else, maybe come along every once in a while and help out. I've done more than I hoped to do. But I find it sad that there is not a million dollars somewhere to save the 220 species that are threatened with extinction. Many of these species just don't have time to wait for the help they need to stay alive."

In the winter of 2021, I inquired to see if Perlman had indeed retired. There was nothing in the news or on social media to indicate that he had. When John Letman, an editor at National Tropical Botanical Garden confirmed it, I was a little surprised, because I had a difficult time believing he would have gone through with it. The lack of fanfare around his exit was testament to the modesty of a man who had no need to be given a "shout-out" for having his name attached to so many plants that he discovered. It reminded me once again of John Bartram, who was a Quaker, and who, like many Quakers other than Richard Nixon, was modest in the extreme, even when Carolus Linnaeus, famous for naming, ranking, and classifying organisms, described him as "the greatest natural botanist in the world."

Chapter 5

Ash Meadows, Ancient Bogs, and Desert Fens

In 1949, American composer Ferde Grofé wrote a short symphonic suite that immortalized the real-life westward journey of William Lewis Manly and John Rogers, two young guides who traveled through the Mojave Desert with destitute families that had gotten hopelessly stranded on route to the California Gold Rush a hundred years earlier. The score to *Death Valley Suite* was inspired in part by the journals that Manley and others left behind. The music was intended to evoke images of sandstorms, desert oases, and Native Americans attacking an oxen-powered wagon train. "Oh! Susanna," among the most popular American songs ever written, was mixed into one joyous part of the medley.

The symphony, along with square dancers and an accompanying choral pageant, was performed in a natural amphitheater in Death Valley at Desolation Canyon by the 86-member Hollywood Bowl Orchestra and the 100-voice Redlands Choral Group.[1] It was "Woodstock in the desert," nearly a half century before the rock festival in Upstate New York made world news. Radio stations, auto clubs, and the California government beckoned people to come, and to bring water, food, and camping gear. People were asked to register their intent beforehand.

Not everyone did. An unexpectedly large crowd of 65,000—one estimate was as high as 100,000—drove in from all parts of the West to hear and see it. Traffic jams prevented most of them from getting there on time.[2] The actor Jimmy Stewart, fresh off two big movie hits—*It's A Wonderful Life* and Alfred Hitchcock's *Rope*—was stuck at the ranch where he was staying. In order to get him to the site on time, the army ferried him across the desert in a jeep.

With Grofé holding the baton, Stewart narrating the story, and various celebrities such as Edgar Buchanan (*Shane*, *Tombstone*, *Petticoat Junction*, *Green Acres*) giving voice to the traveling pioneers, planes flew overhead to see what was happening below. No one had seen anything quite like this free spectacle in the desert, although even more bizarre developments, such as plans for a 34,000-unit trailer park, replete with sportfishing and water-skiing opportunities, and the United States Supreme Court deciding against cattle barons in favor of water rights for the world's most endangered fish, would later come into play in the Mojave.

So many people came to Death Valley that December that there was not enough fuel available to get them all home, nor food to sustain those who got stranded. A local inn and ranch manager who was involved in organizing the event ended up bringing in chefs to make soup. Eventually, everyone went home happy.

Some of the history of this concert is lost or forgotten. In a reissue of the symphony, Naxos, the venerable classical music recording company, wrongly dates the debut of the *Death Valley Suite* as 1957. It makes no mention of the desert show, Grofé conducting, or Jimmy Stewart and other actors being part of it. A live recording was to follow, but there is no evidence of its release.

What we do know is that Paiutes and Timbisha Shoshone did not attack the wagon trains, as they did in the pageant. Rather, in the real story, there were no shoot-outs, but there were arrows shot into cattle and oxen at night, in retaliation for the famished forty-niners stealing pumpkins, corn, and squash that the Native Americans had grown near a peaty fen nearby. "Oh! Susanna" was sung by some of the forty-niners. But instead of "I came from Alabama with a banjo on my knee," they sang: "I came from Salem City with my washpan on my knee. I'm going to California, the gold dust for to see."

Square dancing, and the rising, thunderous medley of Grofé's trail music conveyed the utter despair that sank the spirits of many of the forty-niners for most of their desert journey. Manley admitted to weeping openly and often. In one heart-wrenching moment recalled by Manley after he and Rogers returned from a scouting trip, he describes two mothers in dire distress.

"The four children were crying for water, but there was not a drop to give them, and none could be reached before sometime the next day," Manly recalled. "The mothers were nearly crazy, for they expected the children would choke with thirst and die in their arms, and would rather perish themselves than suffer the agony of seeing their little ones gasp and slowly die." In another scene, two forty-niners were so thirsty that they killed and drank the blood of an ox that seemed to be on its last legs. When they finished drinking, they only wished that there had been more blood to consume.

The one thing that kept them all alive, besides the dried beans they had, the oxen they slaughtered, and the squash and pumpkins they stole, was fossil water—rain, snow, and melting glacial ice that had flowed out of the mountains 8,000–15,000 years earlier into the lakes, rivers, and peaty wetlands that once covered much of the Mojave below.

As I stood in the largest remaining oasis in the Mojave Desert, where those forty-niners stopped to fill up their water barrels, bathe, and allow their wasted cattle, burros, and oxen to graze, it was easy to imagine the struggles they endured. Here, in what is now Ash Meadows National Wildlife Refuge, geological forces have cut a giant gash into the earth. The scar is so long and deep that parts of the desert, including Death Valley, which is about thirty miles downslope of Ash Meadows, are well below sea level.

Ash Meadows gets its name from the velvet ash trees that were much more prominent back in 1849 than they are now—thanks, in part, to salt cedar, which is killing them. It's an invasive species that thrives in the desert because it outcompetes native species for the moisture that comes from streams and groundwater. It's akin to what the thirsty strawberry guava does to the hydrology that native plants in the Alaka'i Swamp were once accustomed to.

There is a great deal of salt, saltgrass, sagebrush, and screwbean mesquite, and a lot of scorpions, sidewinders (rattlesnakes), tarantulas, and tarantula hawks (a species of wasp), just as one would expect in a desert. Relentless, lip-cracking winds routinely kick up salt-crusted sand into giant dust devils that dance like desert roadrunners, across the whitewashed, calcined flats. Up until their extirpation from this area in the 1920s, Mexican gray wolves howled at the full moon while pausing in their hunt for mule deer and desert bighorn sheep.[3] In their absence, coyotes do the yipping now while preying on smaller game like jackrabbits and cottontails.

In a place where temperatures rise well above 100 degrees Fahrenheit in summer, water is naturally a commodity in short supply. Just three inches of rain fall annually in Ash Meadows—even less in neighboring Death Valley. Most of it comes in cloudbursts that cause flash floods. No surprise. then, that director George Lucas chose this part of the world as one of a number of settings for Tatooine, the desolate desert planet in the *Star Wars* films.

I had come here not so much for the desert experience as to see the remnants of that vast lake, river, and wetland ecosystem that once dominated this landscape when a warming climate began turning the snowpack and glacier-fed lakes and rivers into fens and marshes, and then the wetlands into tufa towers, swelling soils, xeric shrublands, desiccated playas, and rubble as water diversions and regional groundwater pumping came into play in the twentieth century.

What I did not fully expect to see in one of the hottest, driest places in North America were other similarities to the Alaka'i Swamp. Here, as in Alaka'i, there is peat and there are wetlands, though not much of either. The Ash Meadow fens, according to peatland ecologist David Cooper, likely had an abundance of sawgrass (*Cladium californium)* similar to the Everglades species. It also has bulrush (*Scirpus americanus*), as well as monocots such as black bog rush (*Schoenus nigricans*).

There are many plants, insects, and animals in Ash Meadows that are found nowhere else in the world. Where Alaka'i has giant drag-

Native Americans settled around spring pools and fens in what is now the Ash Meadows National Wildlife refuge in the Mojave desert. (Photo: Edward Struzik.)

onflies (pinao), luminescent lime-green crab spiders, predatory lacewings that act like ground beetles, the fabulous green sphinx moth, and honeycreepers flitting from ohi'a treetops to lapalapa trees, Ash Meadows has blazing stars, gumplants, lady's tresses, several snails species, a montane vole, and four species of fish—none of which are found anywhere else in the world. Only one other place on the continent—Mexico's Quatro Ciénegas ("Four Marshes"), a wetland in the Chihuahuan Desert—has more endemic species.

Like the endemic species at Quatro Ciénegas, most of the plants and animals at Ash Meadows are linked in some way to the springs and seeps that are connected to a massive underground aquifer flowing from the Yucca Mountains to the north and, to a lesser extent, from the Spring Mountains to the east.[4] By one account, there are 11,000 gallons of water flowing into the Meadows every minute.[5] One of the springs feeds Carson's Slough, site of a 2,000-acre peaty wetland that was nearly drained and mined out in the 1960s. Another is Devil's Hole, a limestone pool of clear, close-to-body-temperature water hundreds of feet deep above a geothermal abyss.

Divers have tried and failed to find the bottom of the fault-fractured pool. Two who snuck under a protective fence so that they could dive into the abyss at Devil's Hole never resurfaced, their bodies sinking into a subterranean water system that may extend into Mexico. Evidence of this occurred on March 20, 2012, when a mini-tsunami in Devil's Hole corresponded to a 7.3 magnitude earthquake in Oaxaca.

This is where one of thirty species of the critically endangered pupfish in the Mojave continues to dwell in splendid isolation, thanks to a 1976 US Supreme Court decision that sided with protection of pupfish over a ranching company's desire to pump out so much groundwater that the fish would no longer have the food and water they need to survive. An inch long, and iridescent blue if the fish is male, the Devil's Hole pupfish looks, as all pupfish do, as if it would be more at home in a coral reef or in someone's aquarium.

Water is key to both places, as it is to the maintenance of all peat and wetland ecosystems. In Alaka'i, the water comes as rain that falls almost daily. At Devil's Hole and the other springs that well up at Ash Meadows, it's fossil water that long ago seeped into an aquifer such as the one below the Nevada nuclear test site, where fossil water starts its 10,000- to 15,000-year journey to Ash Meadows, ninety miles to the southeast.[6,7]

"It takes that long for rain and snow that melts on those mountains all around us to get from there to here," refuge manager Corey Lee told me as he led me on a walk to Crystal Springs where another subspecies of pupfish, the Ash Meadows Amargosa pupfish, lives. "What we have here is an island of water in the middle of a very big desert. It's all that is left of that vast wetland that existed when the climate was much wetter and cooler than it is today."

Lee arrived here five years earlier by way of Florida, where he was a biologist working in the hot, humid environment of Lake Okeechobee, that large shallow body of water that sends far too many agriculture-related nutrients into the Florida Everglades. Before that, he was a graduate student at the University of Oklahoma conducting research on trophy blue catfish. Unlike his predecessor who lived in a trailer behind a ramshackle visitor's center at Ash Meadows, Lee drives ninety minutes from Las Vegas each day to a new facility. It's mostly to accommodate his wife, who works in Las Vegas, and his son, who goes to school there.

Lee allowed that he prefers perennial blue-sky days in a blazing hot desert over the thunderous humid afternoons that routinely spawn tornados in Oklahoma.

"But I really came here for the pupfish," he explained as we watched the fish dart around in the crystal-clear, turquoise-colored pool, feeding on spongy mats of emerald-green algae that seem so alien in a desert environment.

Ichthyologist Carl Hubbs called them pupfish when he began studying them in the 1950s. They reminded him of spunky little puppies chasing each other's tails. Like puppies, they'll eat just about anything.

Pupfish, algae, and some thirty endemic species of plants and animals found at Ash Meadows are relics of those times when glacier-fed lakes, rivers, and wetlands covered the high country. As those glaciers receded, lakes and fens evaporated and eventually dried up. And as the water receded, pupfish and other creatures found refuge in the spring creeks and fault-fractured pools of water that were sustained by groundwater. Like the endemic species of isolated ecosystems such as Quatro Ciénegas in Mexico, the Hawaiian islands, the Galapagos, Dalhousie Springs on the edge of a desert in Australia's Witjira National Park, and the Indonesian island where five new species of songbirds were recently discovered,[8] these plants and animals evolved in isolation and adapted to rapidly changing conditions. What makes pupfish unique is that they evolved in a way that allows them to thrive in water so warm and salty that it would make a catfish go belly-up in seconds.

Crystal-clear water in a hot desert of sand is what drew people to this part of the world for thousands of years. Ancestors of the Southern Paiute and Timbisha Shoshone would come to bathe, drink, and scoop pupfish out of the pools to dry for food. In late summer and fall, they would harvest pinyon nuts and screwbean mesquite seed pods and grind them into flour. (The grinding holes we saw in the rocks at Kings Pool are testament to that.) Like the Tuscarora of the Albemarle Peninsula in North Carolina, they transformed local grapes (*Vitis arizonica*) into beverages. Saltgrass would be used to spice up a meal. Rabbitbrush was for chewing gum. Sticks from ash would be used to make utensils and spears to hunt chuckwalla lizards. Arrows were made from creosote bushes and arrowweed to bring down larger game.

Carson Slough was where they grew grow corn, squash, and other vegetables in moist, peaty soil. These were cooked in clay pots that were coated with the thick syrup that came from desert globemallow.

This was a sacred, haunted, and sometimes terrifying place to be, just as bogs, fens, and swamps were most everywhere. Sacred for the water and food the wetlands offered. Haunted by spirits such as "water babies" rising up from the depths of springs, snatching children that lingered in the pools for too long.[9] Terrifying because of Tso'apittse (carnivorous giants) and Nimerigar (little people) lurking near the springs, ready to pounce on those who strayed too far into the desert.

The forty-niners who passed through knew none of this, of course. Louis Nusbaumer, a German immigrant, was one of those who stand out, because—like William Manley, who lost his way guiding families to California—he left behind an account of his journey through Ash Meadows. Nusbaumer was the grandson of a Swiss-German surgeon who served in the army of Napoleon Bonaparte on the disastrous French campaign across Russia. Nusbaumer's father was also a doctor and expected Louis to be one as well. But his son chose instead to try his hand at farming before failing at that and emigrating to New York.

When an 1849 outbreak of cholera began taking its deadly toll once again on New Yorkers, Nusbaumer headed west with other Germans, promising his wife that he would send for her once he was settled. Along the way, he might have crossed paths with the likes of William T. Coleman and Elias Jackson Baldwin, two men who would make a name for themselves once they arrived in the Golden State.

Instead of joining up with them and others who seemed to know their way, Nusbaumer ended up on the same path that took Manley, Rogers, and others through the desert.

At Ash Meadows, where some of the forty-niners arrived parched, sunburned, and demoralized, Nusbaumer wrote appreciatively of flies, butterflies, and beetles, as well as "magnificent warm water" that was "magical" in appearance (Devil's Hole). It was such an intoxicating experience that he and another man made the long walk back from their camp at Collins Spring a second time on Christmas Eve just for the pleasure of "an extremely refreshing bath."[10]

William Manley recalled that the water at Carson Slough "seemed to be medicated in some way by the way in which it acted on those who drank much of it."[11] (The water may have contained arsenic, which, in small amounts, can cause drowsiness and confusion. The visitor center has a filtration system to screen it out.) Manley, of course, found a way out and nobly returned to rescue those who had stayed behind.

Many of those who came to Ash Meadows in the years that followed were outlaws and outcasts fleeing justice and/or civilization, just as fugitive slaves had done taking refuge in the Great Dismal Swamp for more justifiable reasons. Like those maroons, some outlaws stayed in the desert, adapting to what the desert oasis had to offer.

Men such as Andrew Jackson (Jack) Longstreet represented a peculiarly American folk type that included better-known figures such as Jesse James, Butch Cassidy, and Harry Longabaugh ("the Sundance Kid")—men who defied the established system of justice in favor of a so-called higher form of law. Longstreet was a miner, gunslinger, saloon owner, farmer, and desert horse racer who settled in Ash Meadows around 1890. He was admired by some and despised by others because he stood up for Native Americans when they were denied access to the water they needed for their crops, or when their crops were burned by white settlers hoping to drive them out. Protecting them was his higher form of law.

Longstreet had looks that would have tempted the pen of a modern-day Hollywood script writer. In his youth, he was tall, blond, blue-eyed, and handsome. A cropped ear was the punishment he had received for stealing horses when he was young. The cane he leaned on in old age suggested he had spent too many days bumping along in a saddle and soothing his sore back with moonshine. Legend has it that Longstreet cut notches into his pistol to keep a record of the number of adversaries he had shot.

Longstreet lived out his life in a stone cabin at Roger's Spring with his Paiute wife, Fanny, sensibly integrating the wall of the pool to keep the house cool and supposedly out of the view of any armed posse that might be looking for him.

Aaron Winters was another one of civilization's castaways. He and his Spanish American wife, Rosie, lived on lizards, ducks, mesquite beans, and whatever wild game they could find near the fifteen-foot-square stone hut that Aaron built along a spring that is now called Winter's Hole. According to one visitor, Winters was "a highly respected member of the most exclusive social circle of the desert"—basically an outlaw who shot two, possibly three men before settling in a place that could scarcely support one family. According to the visitor, the only signs of a more civilized life inside the Winters' home were "bottles of Hogan's Magnolia Balm, Felton's Gossamer for the Complexion, and Florida Water—all, alas, empty but still cherished by his wife," who was "little fitted for the privations of the desert."[12]

The Winters' fortunes took a turn for the better when a visitor paid a call, seeking advice on the whereabouts of borax, a desert mineral that was then in high demand. Winters was prospecting for gold and silver at the time, and he knew very well what the man was looking for when it was described to him. But rather than reveal the location, he staked a claim for himself and sold it to William T. Coleman, the Kentuckian who had crossed the desert in 1849 before making a fortune in shipping in California.

There's poetic justice in the fact that the names of many of these characters—Winters and Longstreet, clay miner Johnnie Bradford, merchant-miner Ralph Jacobus "Dad" Fairbanks, farmer and boarding house owner Bob Tubb—are now attached to some of the springs and bogs.

It's a pity that their wives—Rosie Winters, Kitty Tubb, and Celeste (Cettie) Fairbanks—didn't get the credit they deserved. They worked just as hard keeping the farms, boarding houses, and brothels at Ash Meadows going. And each one of them was as colorful as her husband was.

"Shotgun Kitty" was the female version of the American folk type, an Annie Oakley of the desert who never went anywhere without a shotgun. She was a sixteen-year-old runaway when she put an ad in a Philadelphia paper advertising herself as a mail-order bride. When "Big Bob" Tubb took her up on the offer sometime around 1900, she headed west and ended up with him at Ash Meadows. She helped run his brothel and boarding house as well as the 340-acre farm he later worked with water piped in from Crystal Springs. By one account, Kitty was "pretty and pleasingly plump," a poker-playing saloonkeeper who was such a good shot that no one messed with her when she got into a rage. Tough as she was, she had a soft spot for the sick, nursing them back to health for as long as she was needed.[13]

Cettie Fairbanks, on the other hand, was a slip of a woman who ran the family boarding house. Unlike Kitty, her best friend, she was prim and proper and positively Mormon when it came to her views on alcohol. When her husband spent $125 on a 500-gallon barrel of whiskey to sell to patrons at the boarding house, she waited for a quiet moment to turn on the spigot so it would run dry. According to one newspaper account, "one old desert rat dropped to his knees and drunk up what he could from a whisky puddle."[14]

It's also a shame that only one of the springs—Indian Spring—pays homage to the Southern Paiute and Timbisha Shoshone who lived here long before any of the others arrived on the scene. Up until 2000, when the Timbisha Shoshone Homeland Act was put into force, the Shoshone did not participate in refuge management planning.

No one knows how extensive the Ash Meadows peatlands were in those days. The fens may have remained intact had Winters' discovery of borax not set the stage for a century of bizarre and improbable developments that degraded the oasis beyond what had already occurred before the 1880s.

It began with the need for water to quench the thirst of the twenty-mule-train expeditions that Coleman relied on to haul some sixty million tons of borax out of the desert between 1883 and 1889.[15] Borax back then was an ingredient used in ceramics and as a replacement for mercury in the gold-extraction process. When the demand for borax faded, miners turned to silver. To feed the miners, high-risk entrepreneurs like Charles King, a forty-niner who had made a modest fortune as merchant, lumberman, and sheriff in California, brought in 1,300 head of cattle to graze among the cacti, mesquite, and coyotes. In order to provide food for the cattle, alfalfa and corn were grown in moist, peaty soils such as those at Carson Slough.

To keep the miners and ranch hands happy, more boarding houses were built. Ash Meadows became a road stop of sorts for ranchers, prospectors, and occasionally celebrities like Diamond Tooth Lil. (This was Evelyn Hildegard, the notoriously oft-married saloon singer and brothel keeper who passed through and overnighted at "Dad" and Cettie Fairbanks's boarding house. Fairbanks dined with her that night, but Cettie refused to let her daughters anywhere near them.)

All this traffic didn't necessarily spell the end of this peaty oasis. When biologists with the Death Valley Expedition camped at Ash Meadows in 1891, they reported seeing many plants, such as purple loosestrife, growing in "permanently wet boggy soil." A half century later, Carl Hubbs and Robert Miller also saw promise in the land and they successfully lobbied to have Devil's Hole added to the Death Valley National Monument in 1952.

On our hike to the fenland where alfalfa was grown and cattle grazed, we passed through saltgrass, sagebrush, desert mistletoe, and quail brush, the only known host plant for MacNeill's sootywing caterpillars, which are critically imperiled in Nevada and California.

We did not see Ash Meadows sunray, blazing stars, or the Ash Meadows gumweed (*Grindelia fraxinipratensis*), which are found nowhere else in the world. But we did see milk-vetch and desert milkweed, a plant that offers up nectar and pollen to tarantula hawks.

These wondrous, three-inch-long spider wasps are known for the "hill-topping" practice of sitting high on desert plants that can support them while waiting for females to come by. Once the quick business of mating with the male is settled, the female fearlessly hunts down and stings a tarantula spider with a paralyzing venom. She then drags the spider into a desert burrow and lays a single egg into its abdomen. When she closes the burrow's entrance and leaves the paralyzed spider in the dark, the larva hatches and eats the spider from the inside out, studiously avoiding vital organs in order to keep the tarantula alive for as long as possible.

I was happy to learn that it is rare for tarantula hawks to sting people, because the pain, according to entomologist Justin O. Schmidt, the originator of the Schmidt sting pain index, is "blinding" and "shockingly electric."

When we got to the spot we were aiming for, Lee stopped to inspect a screwbean mesquite to see if it was infected by a mysterious disease that has been killing the trees throughout southeastern California, southern Nevada, and parts of Utah, Arizona, New Mexico, western Texas, and northern Mexico. No one knows what is behind the carnage, but like the mysterious pathogen that is wiping out ōhi'a trees in Hawaii, this one is a rapid-fire killer.

"The gnarly seed pods are an important food source not only for Indians but for birds and rodents," Lee told me. "Black-tailed gnatcatchers, loggerhead shrike, verdin, Bewick's wren, and other desert birds also use it for cover. The endangered least Bell's vireo nests on its branches. The presence of mesquite is a sign that there is water close by."

It was weird standing there. There was nothing to help me visualize clover-like fields of alfalfa growing under a scorching hot sun, or cattle grazing peacefully among the ash trees and the cacti. Nor could I imagine that this was once duck-hunting country. So much effort went into digging, diverting, and draining the water. It was hard to fathom why anyone would expend so much time, sweat, and muscle on a desert enterprise that promised so little in return.

And yet there was always someone who saw something in places such as Ash Meadows that few others could see. In the 1960s, it was local rancher George Swink, who decided to drain 2,000 acres of fens to mine them for peat and sell it to local farmers.

After three years of digging, Swink sold the land to Spring Meadows, Inc., a California company that was buying up desert land from the US Bureau of Land Management at bargain-basement prices. In an effort to condition the soil for large-scale agricultural purposes, the company bulldozed sand from nearby dunes into what little remained of the peat. Roads and concrete ditches were built and water courses were diverted to fill up artificial reservoirs. Additional sources of water were pumped out of the ground.

It's hard to believe now, but in 1967 the company was cultivating 4,000 acres of land, grazing 1,800 head of cattle, and employing 80 to 100 people. How much peat was destroyed or degraded is anyone's guess. But it must have been a tremendous amount.

Well before Swink arrived on the scene, Carl Hubbs was already anticipating the damage that entrepreneurs like Swink might do to springs such as Devil's Hole, which was added to the Death Valley Monument in 1952. But being added didn't amount to much in the way of legal protection. The government's monitoring of water levels demonstrated clearly that the more water Spring Meadows pumped out of the ground, the less water there was in Devil's Hole for the pupfish.

Wetlands and nongame fish were not high on anyone's agenda at the time, as the draining of Carson Slough demonstrated. And no one, not the US National Park Service, the US Fish and Wildlife Service, nor the Sierra Club, The Nature Conservancy, or other well-established conservation organizations, expressed any concern about the introduction of bullfrogs, mosquitofish, red swamp crayfish, and other invasive species that played a role in the extinction of Ash Meadows killifish.

The ecological importance of freshwater fish was not a major concern in the 1950s. "They were something to be caught and eaten. . . . Nongame fishes were viewed primarily as competitors for game fish," is the way Edwin Pister, a California Fish and Game biologist, put it when he got interested in what was happening at Ash Meadows.

Pister arrived just at the right time with a new paradigm of thought that followed a disastrous attempt in 1962 to eradicate native non-sport fish along 444 miles of the Green River in southwestern Wyoming and northeastern Utah. In what amounted to the biggest fish-poisoning plan ever launched in the United States, 21,100 gallons of poison (rotenone, an insecticide) was introduced at 55 drip stations upstream of the nearly completed Flaming Gorge Reservoir. The intent was to kill so-called trash fish such as the common carp, introduced to the river in the nineteenth century, before stocking the artificial lake with lake trout and kokanee. Scientific studies, environmental reviews, and permitting were deemed unnecessary. No one considered the fact that many other native species, such as razorback suckers, Colorado pikeminnow, and humpback chub, would die as well.

The carp and other native "trash fish" were successfully poisoned. But the so-called detoxification that was supposed to occur downstream of the reservoir was botched. The die-off was so massive and so painfully obvious to sportfishermen and the public that Interior Secretary Stewart Udall took action to ensure the following "future measures such as this would be reviewed, and everything would be done to prevent the potential loss to the pool of genes of living material is of such significance that this must be a dominant consideration in evaluating the advisability of the total project. I am taking measures to assure that future projects are reviewed to assure that experimental work is taken into consideration, and that possible deleterious effects are evaluated by competent and disinterested parties."[16]

For the pupfish of Ash Meadows, this declaration set the stage for one of the longest, most high-profile environmental battles in North American history.

That battle began in earnest with a simple telephone call in March 1968, a year after the Endangered Species Preservation Act was put into play. US National Park Service naturalist Dwight T. Warren called Pister and told him about the "bad things" that were happening at Ash Meadows. Pister organized a field trip to the area the following week to see firsthand the environmental degradation that was supposedly taking place. He and fellow scientists were horrified by what Spring Meadows was doing by pumping out so much water.

Some of them spent the next year trying to get various agencies to do something about it. When it appeared that no one in the upper echelons of government was interested, Pister organized two meetings with fellow scientists and advocates in Ash Meadows, first in April 1969 and then again the following November, to develop recovery and conservation strategies.

It was the beginning of one of the more unlikely scientific organizations established in the United States. At its inaugural meeting in Death Valley in mid-November 1969, the Desert Fishes Council worked "two days and nights in session and in the field forming battle lines to save from extinction some of the nation's rarest and most endangered life forms—the ancient but little-known desert fishes of the Death Valley Drainage system."[17]

Council members were cognizant of the fact they had little expertise in the legal and regulatory technicalities that were needed to get their initiatives going. What they did know was that there were people in government who could help.

Pister, it turned out, had once worked alongside fisheries biologist

Charles Meacham, who was commissioner of the US Fish and Wildlife Service at the time. With nothing to lose in a battle that was not going the Council's way, Pister telephoned Meacham's office in Washington, DC, hoping to get a sympathetic response. Meachum was out of town. When Meachum returned the call later that day, Pister did his best to describe what was happening in the Amargosa Valley without making it too confusing.

Meacham's response was completely unexpected.

"Which spring is involved, the northeast or northwest?" Meacham asked. The Amargosa Valley, it turned out, had been one of his favorite boyhood haunts.[18]

The call led to the establishment of a pupfish task force and a court injunction preventing Spring Meadows from pumping any more water. As the matter proceeded slowly and incrementally to the Supreme Court, battle lines were drawn between, on one side, land developers and agricultural interests and, on the other, those who supported pupfish and wetland preservation. "Kill the Pupfish" bumper stickers were printed by Nye County commissioner Robert Rudd in response to the "Save the Pupfish" stickers that were produced as a joke by a California assemblyman. Local newspapers sided with Spring Meadows with such zeal that one editor suggested that "an appropriate amount of rotenone dumped into that desert sinkhole would effectively and abruptly halt the federal attempt at usurpation."[19]

The national media was more sympathetic. Thanks in part to the fallout from the Green River fish poisonings, the public was beginning to express concern about other environmental issues such as DDT and the overhunting of charismatic species like the polar bear. The magazine *Cry California* devoted an entire issue to pupfish in March 1970. NBC television followed two months later with a documentary, *Timetable for Disaster*, which was filmed, in part, at Devil's Hole and narrated by Academy Award–winning actor Jack Lemmon. In short order, the *New York Times*, *Wall Street Journal*, *San Francisco Chronicle*, and other major newspapers devoted ink to the issue.

High-level politicians like Stewart Udall and John Gottschalk, director of the Bureau of Sports Fisheries and Wildlife, stepped in. Pitching the case for the government acquiring more land for conservation, Gottschalk told the media that "dogs, poachers, and hurricanes aside, endangered Key deer cannot live in the midst of pizza parlors, and a marsh blacktopped for an airport will not sustain marsh dwellers."[20]

The case ended up in the Supreme Court, where the justices sided with the pupfish.

Leo Cappaert, one of the owners of Spring Meadows, Inc., realized that he had no other option but to sell the land when the judges ruled against him and his family-run company. Inexplicably, and to the dismay of all who were involved, the Fish and Wildlife Service passed on his offer to sell the water rights and about twenty square miles of land at Ash Meadows. Spurned, Cappaert Enterprises sold its interests in 1980 to Preferred Equites, a land development corporation.

I had no reason to doubt Lee when he told me that, in spite of all the publicity and the Supreme Court decision, Preferred Equities forged ahead with a plan to build a new town at Ash Meadows, with 34,000 trailer homes as well as hotels, shopping centers, and an airport. They even had plans to build a reservoir so that the residents could fish and water ski. It reminded me of the "Garden of Eden" scheme that "Dicky" Bolles came up with in 1911 when he sold swampland in the Everglades to people who did not know that the land was underwater.

Preferred Equities may have gotten their way had the amount of water needed for the development not been 368 percent more than the volumes that were being pumped out of the springs.

Sensing an opportunity, The Nature Conservancy reached out to Jack Soules, president of Preferred Equities, to see if he was willing to sell the land. His asking price was five times the value—too high for the Conservancy to seriously consider or afford. But then, a month after Soules declared in April 1982 that nothing was going to stop this development, the most unlikely person stepped in to save the day. Interior Secretary James Watt, the man who leased massive tracts of federal land to coal-mining companies and opened up the outer continental shelf to oil companies, authorized emergency listing of the Ash Meadows pupfish and Ash Meadows speckled dace. Faced with being in violation of the Endangered Species Act, Soules had no choice but to sell.

The Nature Conservancy purchased the land for $5.5 million in 1984 with the understanding that the US Fish and Wildlife Service would buy Ash Meadows and turn it into a national wildlife refuge.

As I headed back to the bright lights of Las Vegas that night, I kept thinking that you can't make this stuff up. But there was more to come: drunks, including a convicted felon, jumping fences and vomiting into Devil's Hole, and a religious group suing the Fish and Wildlife Service for damage done to their property when rewetting efforts in Ash Meadow diverted water away from their sanctuary.

The one even more important thing that struck me most was that the Ice Age is still with us, delivering fossil water from melting snow and glaciers that covered a significant amount of the continent more than 12,000 years ago. This fossil water still flows through to Ash Meadows, as it does in other hot, dry places such as Orange County, California, which was site of one of the biggest peatlands in the state before land speculators began buying up land with so much enthusiasm that five acres of peatland sold for a record $3,000 in 1899.[21]

Like those in Ash Meadows, the more expansive peatlands in Orange Country were ditched and drained and worked over with so much vigor that workers draining a wetland near one ranch were caught off guard when a gush of water erupted from a subterranean lake that was so deep, a twenty-foot pole did not reach bottom. "The drained area was flooded by the eruption," noted the *Los Angeles Times*, "and it will have to be drained again."[22]

The wetlands of Orange County did not last long. But peat got its revenge when it began frequently to catch fire as it dried out. One of those fires famously burned for four years in the 1920s, threatening the health of 100,000 residents who lived downwind of the smoke. Several people suffered burns, including one woman who testified that when she set foot on the burning peat, she sank. "I had to step on my ukulele to get out," she recalled before posing with the musical instrument for a photo in the *Los Angeles Times*.

Long-suffering residents finally convinced authorities to sue the owner of the wetland bog. It was Anita Baldwin, heiress to the wetland that Lucky Baldwin had drained years after he made that wagon-train trip to California.

Chapter 6

Sasquatches of the Swamps

One gram of peat usually contains some 50,000 wind-deposited pollen grains.

— Bryologist Howard Crum, *A Focus on Peatlands and Mosses* (1992)[1]

The Aweme borer (*Papaipema aweme*) is a yellowish-brown moth with an inch and a half wingspan. In the world of lepidopterology—the study of moths and butterflies—it's as flashy as a house sparrow might be in ornithology, but as elusive as the ivory-billed woodpecker before it was finally deemed to be extinct. For a long time, entomologists thought the moth lived in the sand dunes and oak savannahs in southern Manitoba and the Great Lakes region of North America. No one really knew. Up until 2005, only six specimens from four widely scattered locations had ever been found. Many doubted it still existed at all until one was trapped in a peatland fen in the backwoods of Upper Michigan in 2009.

For Wisconsin-based entomologist Kyle Johnson, that was a game-changing moment. His easygoing search for *P. aweme* switched gears into an intense hunt. Instead of focusing on the sand dunes and alvars of Upper Michigan and northern Wisconsin, he and his colleagues put on rubber boots, mosquito jackets, and bug hats and began squishing through dozens of peatland fens. They traveled nearly a thousand miles from the Upper Peninsula of Michigan to eastern Saskatchewan on the Canadian prairies looking for "the beast," as entomologist Chris Schmidt once described it.

Many a night, Johnson slept briefly in his car, in a tent in a roadside ditch, or on a hammock that he hooked up between two gnarly black tamarack trees; the bite of a mosquito on his posterior being preferable to the groundwater that gradually seeped up under the weight of a tent and sleeping pad. Hotels weren't an option because night is when most of the work was done. And anyway, most hotels were simply too far away. In all, he and his colleagues spent 123 nights in buggy peatlands capturing moths at bait stations and netting free-flying adults.

In that eight-year search for *P. aweme*, Johnson and his colleagues recorded 59 specimens, all of them in fens that had a lot of buckbean (*Menyanthes trifoliata*) growing in them. As is sometimes the case with the larvae of moths and butterflies, there is a specific plant they are attached to—the monarch butterfly's reliance on milkweed being one of the more notable. The showy white flowers of the sacred datura in Ash Meadows is another; hawk moths have long tongues that allow them to get the nectar from the trumpet-like blossoms.

More obscure is the recent discovery of a new species of moth that was reared from showy lady's slipper orchids growing in bogs in eastern Ontario and southwestern Quebec,[2] and *Monopis jussii*, a Tineid moth that was found in the fall of 2020 living in the nests of boreal owls in Finland.[3]

When I met up with Johnson on a warm summer day, he was still hunting, almost certain that the range of *Papaipema aweme* extended well beyond eastern Saskatchewan into the fens of north-central Alberta.

"The geography of these early recordings suggested, and I mean only suggested, that *P. aweme* was a denizen of dry habitats such as sand dunes and oak savannah," he told me. "Many entomologists didn't think to look elsewhere because a lot of them didn't believe that many moths could be peatland specialists."

We were in the middle of Clyde Fen at the time, which is five miles from the small nowhere town of Clyde in a tiny wetland in Alberta.

No one in any of the conservation-related departments of government knew about the Clyde fen until the early 1980s, when Canadian Forest Service scientist Derek Johnson alerted them to the fact that it was possibly the most southerly home of the carnivorous pitcher plant in the province and a critical habitat for bog adder's-mouth, an extremely rare orchid that has a chameleon-like ability to blend into mounds of moss.

With us that day was Greg Pohl, a Canadian Forest Service entomologist who had taken the day off from his government job to do what he does in much of his spare time. It was Pohl who invited me to come along when he learned that Johnson was on contract with the Canadian Wildlife Service on a summer-long mission to find *P. Aweme* in this part of the world.

Pohl is tall, thin, and thoughtful-looking in the manner of a person who understands the path to discovery. He himself has found two new species of moths, both of which were masquerading as other native species. His enthusiasm for lepidoptery reaches a zenith each June, when he hosts a Solstice Party in his backyard. Pohl puts out a white blanket and a glow light to see what kind of moths show up when the sun descends on the longest day of the year.

Pohl was wearing a T-shirt that had a sketch of *P. Aweme* on the back and, on the front, the message "*Extinct? I think not.*" He warned me on the drive into the fen that I might find Johnson to be a little nerdy. He himself didn't know what to expect. "Even entomologists don't know how strange or nerdy they are until you put two of them in a room," he explained. "When one comes out and says, "'Boy, that person is strange,' you have your answer."

At first glance, there was nothing strange about Johnson. To my mind's eye, he fit the stereotypical image of a butterfly or moth hunter. He was 34 years old, wearing rubber boots and a brown boonie hat that he pulled down around his ears to protect himself from the hot sun. His beige pants blended in well with a similarly colored T-shirt and a vest that was only slightly darker. His wire-rimmed glasses were inconspicuous. He was stout, and as physically fit as one might expect a person who spends so much time slogging through bogs and fens to be.

But once he got talking, I realized that he was maybe just as eccentric as entomologist Norman Criddle probably was when he was catching moths and butterflies on the prairies. Criddle was the first person to identify the Aweme borer moth and the last person to see the Rocky Mountain locusts that once swarmed the Great Plains of North

America in the billions before mysteriously disappearing in the 1900s. It was Criddle who came up with the formula to control locusts. It included horse manure, salt, and Paris green, a toxic mixture of copper acetate and arsenic trioxide. Criddle fed it to locusts that he trapped and kept on his farm in Aweme, Manitoba, in the early 1900s. They were the last live ones that anyone had seen.

Criddle, I learned from Pohl that day, is someone whom most everyone with a passion for lepidoptery in North America reveres. Many go on pilgrimages to visit the family homestead, which was bought up by the Manitoba government and turned into a heritage park. The designation is a testament not only to Norman, but also to his many brothers and sisters, who were just as accomplished.

Criddle inherited his eccentricity from his father Percy, an English wine merchant whose life was by all accounts complicated. Percy Criddle had a wife and mistress, neither of whom was aware of the other until they began producing a passel of children as Criddle's business fortunes were waning in Britain. Percy decided to move them all by boat, train, and steamer to Brandon, Manitoba, and then to Aweme, a remote backwater north of the Assiniboine River on the Canadian prairies in 1882. The mistress and four children traveled steerage class while Criddle, his wife, and their four children lodged in more comfortable quarters.

This one big family lived and starved in tents for the first half year because they arrived too late to plant crops. As the summer and early autumn months passed, Criddle and the older boys sawed logs in the mosquito-infested forests to build a cabin. Neighbours took it upon themselves to help, likely out of goodwill, and possibly with the promise of being paid to get the Criddles into something warmer by Christmas. Criddle proved to be a terrible businessman, though his mother, an accomplished English artist, was wealthy.

Criddle was a wonderful entertainer, even if he was a snobbish one. He sang classically and, by some accounts, poorly. He put on concerts for farmers who were no doubt bewildered but happy to be catered to in unfamiliar and ostentatious circumstances. But he was a terrible farmer, choosing to wander off and catch butterflies rather than tend to the sandy fields that failed to produce good crops.

Norman was just seven when this double family settled on the farm, and ten when he took his turn plowing fields while also helping his brothers and sisters cut grass, on their hands and knees with scissors, in order to maintain the family's tennis court. (Scythes couldn't crop the grass low enough to the ground.) His father, deeply absorbed in meteorological, astronomical, and biological observations, did not have much patience with him, or the other children, and showed no interest in sending any of them to school.

Being outdoors as often as he could, Norman developed a keen interest in the natural world. In 1912, a chance meeting with James Fletcher, a government entomologist and botanist, resulted in a series of assignments and an offer to work for the Manitoba government as an entomologist. The high point of his career arrived when the family, with the help of some government funding, built him a lab on the farm in Aweme to study white grubs, Hessian fly, and wheat stem sawfly, which were at that time major problems in Manitoba.

The follow-up to Criddle's discovery of *P. aweme* underscores one of the challenges to getting a better understanding of peatland ecosystems. Benchmark studies of insects are rare. There is no standard in monitoring and inventory methods.[4] Everyone assumed thereafter that it was dry, sandy habitat that the moth preferred, not the wet, boggy ones that make surveys a nightmare.

Subsequent discoveries of *P. aweme* proved to be elusive. On August 13, 1925, a man by the name of Sherman Moore found a single specimen on a boat anchored off Beaver Island, Michigan. On August 7, 1932, another single specimen was taken in Rochester, New York. On August 15, 1936, a moth was netted by an unknown collector on the sandy shores of Lake Huron in Grand Bend, Ontario.[5] Sixty-nine years later, botanist John Morton found one near Pike Lake on Manitoulin Island, Ontario.

"It was a complete enigma," said Johnson. "The geography of these early recordings suggested, and I mean only suggested, that *P. aweme* was a denizen of dry habitats such as sand dunes and oak savannah."

Johnson drives, on average, 20,000–30,000 miles each year looking for moths and butterflies. He estimates that he's explored more than a thousand fens and bogs. "Fens and bogs are my favorite places to explore because there is so much to discover," he told me. "They're soft and spongy, and filled with life—a reminder, I think, of the wonderful times I spent exploring a fen near my grandparents' place in Wisconsin."

Once, when Johnson had driven to Alaska in pursuit of moths and butterflies, he was confronted by a grizzly bear while chasing down an alpine butterfly. "That was a heartstopper," he said. "It sure got the blood pumping."

As the afternoon wore on that day at Clyde, I was impressed by the number of pitcher plants and sundews there were in this fen, shabby as it seemed to be with a road running through it, a large part of it burned, and farms and a gravel pit surrounding it. I was secretly thrilled when I thought I found an extremely rare bog adder's-mouth orchid, but I learned later from orchid specialist Ben Rostron that it was *Platanthera aquilonis*, the north wind bog orchid, which is much more common. The challenge in finding the bog adder's-mouth, according to Rostron (who travels the world hunting for orchids), is that it is not only rare, but also difficult to detect even when you know it is there growing among the mosses.

The real treasure in Clyde, according to Rostron, is *Liparis loeselii*, the wide-lipped orchid that is not supposed to grow in Alberta. It grows in fens, bogs, and dune slacks. It has yellow flowers and glossy yellow-green leaves. "It is there," he assured me. "You just need to know where and when to look."

Had I known, I would have taken the time. But the time had begun to tick down as thunder clouds loomed in the western sky. Several flashes of lightning and the distant rumbling of thunder didn't seem to unnerve Johnson as he and Pohl hovered over a netted moth that later proved to be *Crambus lyonsellus*, a species that has never been found anywhere west of Manitoba. It was this moth that turned out to be the most common one found that afternoon and well into the evening.

I saw a lot of moths being captured, and many signs of larvae chewing up plants. But I wasn't there for the full accounting, because Pohl and I eventually made a run to the car knowing that we were safer there than standing in a couple of inches of water during a spectacular thunderstorm. The last I saw of Johnson as we were driving away in the deluge, he was still out there with net in hand, focused more on what was on the ground than on the light show and the thunderclaps coming from above.

Weeks after Johnson completed his survey of fens in Alberta, he sent a note informing me that he had not found *P. aweme* in the Clyde fen or in any of the other fens he explored over the course of the next week. He reluctantly allowed that maybe, just maybe, *P. aweme* may not dwell farther west than eastern Saskatchewan.

"But that's no reason to stop looking," he said cryptically.

I didn't give this much thought until the following spring, when entomologist Chris Schmidt sent out a note on social media informing me and others that he found *P. aweme* in the Richmond fen just outside of Ottawa, Ontario, 300 miles from the nearest known populations. The larvae were feeding on buckbean, just like other specimens in Michigan and the Canadian prairies. "Good news indeed for this legendary beast," wrote Schmidt. "Maybe it will turn up in Quebec."

I first met Schmidt when he was a student collecting moths in a fen not far from where I live in western Canada. I had hoped to meet up with him in the 7,700-year-old Mer Bleu bog east of Ottawa before the COVID-19 pandemic suspended my cross-continental travels. Unlike Mer Bleu, which covers nearly 13 square miles, the Wagner fen is so small (one square mile) and inconspicuous that people driving by on route to the Rocky Mountains would not know it was there if it weren't for a sign on the road vaguely signaling its presence. Located along the outskirts of the city of Edmonton in Alberta, Canada, it used to be known as a bog before scientists declared it a fen. It is now known as Wagner Natural Area.

On any given day, the small parking lot is empty. Visitors like Ben Rostron, who comes regularly, go in late spring when there are marsh marigolds, shooting stars, lady slippers, and many orchids in bloom.

I had heard of Wagner's fen from university scientists like Bill Fuller who sent their students there to map out orchids, collect butterflies and moths, gauge groundwater flows, and conduct inventories of the birds and animals that were present. Fuller, now deceased, was the scientist who helped solve the mystery of where the world's last remaining whooping cranes nested. He found them in the fens of Wood Buffalo

National Park in northeastern Alberta and the southern Northwest Territories.

I had assumed that the scientific interest in Wagner was related to its close proximity to the University of Alberta, which has a population of 38,000 students, many of whom are enrolled in science programs. It never occurred to me that there was anything particularly special about this place until entomologist Jens Roland introduced me to the elusive world of butterflies with his efforts to track them with tiny transmitters. When Roland saw how impressed I was with that novel experiment, he invited me to tag along with a student who was assigned the task of capturing and cataloguing moths in Wagner. Chris Schmidt was the student.

A fen or a bog at night is an entirely different place than one illuminated by sunshine. Beneath the glow of a gibbous moon, the starry dance of thousands of fireflies played itself out around us that night as we hiked through a thick stand of black spruce, tamarack, lodgepole pine, birch, and aspen. With mist thickening around us, I could hear but not see the big barred owl hooting in the distance. But I did notice the fresh tracks of a moose that had passed through earlier in the day after rain turned the ground into mud. Bears and lynx used to visit, but they are now rarely seen because of the industrial developments that have all but closed wilderness pathways into the area.

Once we reached the clearing that Schmidt was aiming for, he pulled out a white, king-sized bedsheet from his pack and draped it over two tent poles. He then lit it up with a high-powered, battery-operated flashlight that caught the attention of a couple of screeching bats and what I suspected was a flying squirrel whooshing from tree to tree. Primitive as the blanket strategy appeared to be to me, it was less messy, I was told, than sugaring—a moth-trapping technique that involves smearing a pungent brew of rotten bananas, molasses, brown sugar, and beer onto the trunk of a tree.

Most moths are nocturnal. One notable exception is the hummingbird clearwing, which is often mistaken for a hummingbird because it is big and hovers in front of flowers as it feeds. Moths tend to be less flashy than this, which may be why they don't get the respect that butterflies do.

That's a shame because, next to butterflies, moths offer us the deepest and most detailed record of how and why global insect populations are in decline. They are more ecologically diversified, and they are fifteen times more taxonomically diverse than butterflies.[6] The 124 moth families, which include 160,000 known species, are so poorly known that the majority of species have not even been discovered or described. It's an inexcusable slight for a creature that has a fossil record dating back 190 million.

The evolutionary diversity of moths coincided with that of flowering plants 65–145 million years ago when dinosaurs dominated the world. Moths are not a formal taxonomic group. Instead, they are divided into two groups—the larger macromoths and smaller micromoths, which tend to represent the more primitive side of the family tree. The simplest way of determining the difference between a moth and butterfly is to look at their antennae. If the tips of antennae are thickened or club-like, it's a butterfly. If the antennae are feathered or thread-like, it's a moth.

The first insects to show up that night were not moths but caddisflies, June bugs, and handful of edgy yellowjackets that had not yet got the message that it was dark. To add some warmth to what was turning out to be a chilly night, Schmidt offered a sip of whisky that he had brought along.

A giant tiger moth eventually landed on the sheet. When an even larger five-inch long Polyphemus moth (*Antheraea polyphemus*) followed, I was startled by its breadth, beauty, and size. It was a giant, suitably named after the cyclops in Greek mythology, which had a single round eye in the middle of its forehead. This moth has two stunning peepers on its wings, perhaps designed by nature to give predators pause. Its most formidable enemy is a fly (*Compsilura concinnata*) that was introduced to North America in 1906 to control the invasive gypsy moth. At the time, no one foresaw that it would be capable of parasitizing 150 species of moths and butterflies that are native to North America.

The timing of a moth's arrival is dependent on the species and the nature of the trapping site. The Luna moth, for example, is aptly named because it uses moonlight, as some other moths do, to guide it on its nocturnal flights between midnight and 3:00 a.m. It is a giant so rare and elusive in Alberta that, according to Pohl, it has a sasquatch-like reputation. If there are fashion plates in the plain world of moths, the Luna is a showboat, more stunning and Amazonian looking than the Polyphemus, and an equal match, I think, for the Madagascan sunset moth, which is often described as the most impressively appealing of all lepidopterans.

The only Luna moth I've ever seen was one I spotted along the edge of a bog in Lake Huron's Georgian Bay. The lime-green gossamer wings are bound by pink-purple seams that give it a fairy-like appearance. The tiny Playboy bunny antennae and two big Picasso-like eyes offer, along with sexy-smelling pheromones, an irresistible come-hither package. The larvae aren't nearly as enchanting. They make clicking sounds before spitting vomit on predators that get too close.

Schmidt told me that he had collected 220 species of moths the summer before and expected hundreds more would be found in the years ahead.

I had trouble wrapping my mind around these numbers for a fen as small as this one at Wagner—until I spent another day there with Bert Finnamore. Finnamore at the time was curator of entomology at the Provincial Museum of Alberta. He had first visited Wagner back in 1985 when he was searching for sites to conduct a study of arthropods—invertebrates that, like spiders and ants, have no spine and bodies made up of different segments.

Wagner seemed like a good possibility because it was close to Edmonton and was protected by legislation, making it suitable for long-term monitoring. He also thought that the initial survey would be fast and simple.

"Boy, was I in for a surprise," he told me. "Of the estimated 1.5 million specimens we collected, there were 2,185 species of arthropods, contributing to a total of 2,909 known species in the peatlands. We had nine people working on the collection. We never did finish cataloguing them all. Only one site, Wicken Fen in England, had fauna that exceeded what we found in Wagner."

Wagner cuts through a gently sloping coniferous swamp where groundwater from adjacent uplands passes through underlying deposits of glacial sand and gravel as well as peat before percolating back to the surface. I learned this from Ben Rostron, who is a petroleum hydrogeologist at the University of Alberta when he is not hunting for orchids. Rostron took me on a very different tour of Wagner one day to describe the plumbing system that nurtures the fragile orchids that grow there.

Calcium carbonate, typically found in glacial water, eventually precipitates as a white paste called marl. The mineral-rich water is extremely alkaline and flows year-round in Wagner at a temperature of about 4 degrees Celsius (39°F). It is what moisture-loving, frost-sensitive plants such as arrow-leaved crowfoot, bog violets, shooting stars, and fringed gentian need if they are to get through minus-25-degree-Celsius (−13°F) temperatures that are common during an Alberta winter.

Orchids are Wagner's biggest attraction. Of the 26 species native to Alberta and most of the prairie provinces, 16 have been found in Wagner. Among the most enchanting, I think, is the yellow lady's slipper, whose flowers look so much like a miniature wooden shoe or moccasin, it is easy to imagine a fairy cobbling it. It's a deceptive plant, designed by nature to dupe a bee into thinking that it has nectar when it has none to give. With pollen being the only reward, the bee continues on after entering the slipper, unaware that the orchid is designed to strip it of some of the pollen it may have carried from a compatible plant.

By contrast, the flower of the tiny bog adder's-mouth that is also found at Wagner is an aesthetic disappointment, mainly because it is so small. It requires a macro lens, or a poet like Theodore Roethke, to magnify its charms.

They lean over the path,
Adder-mouthed,
Swaying close to the face,
Coming out, soft and deceptive,
Limp and damp, delicate as a young bird's tongue;
Their fluttery fledgling lips
Move slowly,
Drawing in the warm air.
And at night,

The faint moon falling through whitewashed glass,
The heat going down
So their musky smell comes even stronger,
Drifting down from their mossy cradles:
So many devouring infants!
Soft luminescent fingers,
Lips neither dead nor alive,
Loose ghostly mouths
Breathing.[7]

Bog adder's-mouth is at home in fens and bogs because of fungus-loving gnats that do well grazing on slime molds growing on the surface of decaying logs. *Phronia digitata* is the gnat that pollinates this orchid.

The first threat to Wagner came, as was often the case in the nineteenth and twentieth centuries, from farmers like Frederick Wagner, a German immigrant who purchased the site from the Canadian Pacific Railway (CPR) in 1926. His plans to drain, log, and pasture the land got waylaid when he learned that the CPR was planning to mine the marl to produce cement. They claimed to have never given up the subsurface rights to the area.

Wagner would have none of it. He hired Edmonton lawyer William Morrow to take his case to court. Morrow was young at the time, but smart and savvy enough to eventually turn himself into the most famous itinerant judge of the Arctic. He won the case for Wagner, convincing the presiding judge that marl was not a subsurface mineral.

Wagner never did get very far with his plans to turn part of the fen into a dairy farm. While he managed to drain and clear enough of the damp forest to create a pasture for cows, there was something in the fen's ecosystem—possibly seaside arrowgrass—that tainted the cow's milk. Seaside arrowgrass contains cyanogenic glycoside, which can release hydrogen cyanide (HCN, or prussic acid) during mastication by animals. The concentration of toxic chemicals increases during times of moisture depletion.[8]

William Wagner had his own troubles with outsiders after his father died and left him the land in 1946. The inheritance coincided with the huge discovery of crude oil in Leduc, twenty miles distant. Within a few months of that gusher, companies like Shell and Imperial Oil were all over the area drilling for oil. All they got for their efforts around Wagner's fen, however, was salty water that gushed up and nearly flooded out the Wagner heir. Only a direct appeal to the government persuaded the companies to plug the boreholes with cement.

Wagner apparently saw nothing unusual in the number of moose, bears, lynx, owls, and birds that were a common sight on his land. When naturalist and filmmaker Edgar Jones asked Wagner if he could do some filming on his property, he saw no problem in letting him do so. And when Jones later asked him not to cut down trees where Bonaparte's gulls were nesting, he agreed. Like Napoleon himself, the Bonaparte is small in stature. (It was named for Charles Lucien Bonaparte, the emperor's nephew and an important ornithologist.) Unlike most gulls, it nests in trees, and it never visits garbage dumps.

Edgar Jones was a daredevil pilot who came home from the Second World War with a Distinguished Silver Cross. He worked as a bush pilot in northern Alberta and the Arctic, where he distinguished himself once again by saving the wife and daughter of a trapper who had gone missing. (Forty-one years later, the young girl he saved showed up at an art show to thank him).

A year before Jones died, he told me how he asked the farmer if he would be interested in turning the fen into a protected natural area. Wagner allowed that he was open to the idea, so long as it wasn't going to be an act of charity.

Jones, his long-time friend Bill Morgan, University of Alberta scientists, and local landowners formed the nonprofit Alberta Wildlife Foundation which they used to obtain a one-year option on the land. Within that year, The Nature Conservancy and an anonymous donor came up with the money needed to meet Wagner's price.

In 1975, administration of the Wagner Natural Area was turned over to the Alberta government on the assumption that they had the resources to better manage the fen. But having the government as a partner proved to be problematic when the transportation department came up with plans to build a road along the eastern boundary of the fen.

Everyone with a connection to Wagner knew by then that something dramatic needed to be done to relocate the road development. Peatland scientists like Dale Vitt, naturalists like Terry Thormin, botanist Patsy Cotterill, landowner Alice Hendry, and a sympathetic government biologist, Peter Lee, gathered in Vitt's basement one night to figure out a plan and create a society to protect the fen.

The idea, Lee told me, was to come up with something so big and splashy that the government he worked for would have to back off from building the road as planned. A video was produced. But the silver bullet in this case, he said, was proving that rare orchids like the bog adder's-mouth orchid grew in Wagner.

Ecologist Matt Fairbarns accepted a modest $500 offer to find the orchid. He spent three days on his hands and knees before he finally found one. No one in government knew how small and underwhelming the plant was. But its rarity was enough to get the Transportation Department to relocate the road. Someone in government—no one knows who—paid tribute to the campaign by having the image of an orchid imprinted into the cement of the road's overpass.

While Wagner continues to be a refuge for those orchids as well other plants and animals, most everyone recognizes that the days when Terry Thormin accounted for thirteen saw-whet owls, eight boreal owls, two great horned owls, and two long-eared owls in one night are long gone.

It's quite possible, though, that the invertebrate diversity that Wagner is famous for is still intact. The fame is well earned and so well recognized that that Dutch scientist Hans Joosten highlighted the importance of Wagner in a report he did on peatlands for the United Nations.[9]

The last time I hiked through Wagner, I convinced naturalist John Acorn to come along and show me what he sees when he goes there. If there was some angle I was missing, he would be the one to point it out. Acorn is an author, educator, university lecturer, and host of a Discovery Channel show on nature that had a long successful run. John was more than happy to come along, because it gave him the opportunity to look for a butterfly he had not seen at Wagner since the 1980s.

Saw-whet owl chicks in Wagner fen, where two-thirds of western Canada's orchids can be found. Only one site, Wicken Fen in England, had a variety of fauna that exceeded what was found in Wagner when it was surveyed in the 1980s. (Photo: Edward Struzik.)

The cranberry blue butterfly that Acorn was hoping to find is one that entomologists thought they knew a lot about until Acorn, scientist Scott Nielsen, and Federico Riva, an entomologist at the University of Alberta, discovered fourteen unreported populations in the peatland forests of Alberta's Wood Buffalo Region.[10] The butterfly seems to do well in peatlands bisected by seismic lines, according Riva, but it does poorly in oil and gas well pad areas that cover much of northern Alberta.

Not a minute into the hike, a click beetle popped into view. When Acorn poked it, the beetle clicked. "They make that sound by flexing their thorax so that they can jump from the back," he said. "All the energy needed to ward off a predator is released in an instant. It's cool. Presumably it's enough to give a bird a shock."

There was no else on the trail other than a man who came to see ducks. He looked disappointed when we told him that forested fens like this were not noted for waterfowl but instead for showy plants like the cluster of marsh marigolds we were admiring at the time. He didn't seem to care.

As the man marched off unhappily, Acorn began connecting dots for me.

The hover flies we saw along the trail are true flies, even though they look like small bees or wasps. They are valuable in that their larvae feed on aphids, thrips, scale insects, and caterpillars. (The adults feed on pollen and nectar.) They also pollinate the roundleaf orchid (*Galearis rotundifolia*), which grows in Wagner.

The taiga alpine butterfly would be extinct if ecosystems such as this disappeared, because they are black spruce peatland specialists. "The fact that we still haven't figured out which plants their caterpillars eat is another humbling peatland fact," said Acorn.

Acorn once doubted that the eastern forktail damselfly dwelled at Wagner even when a colleague reported seeing it. "I had published a bogus record of the eastern forktail and was mortified when I discovered late that it was a prairie bluet. I figured she had made the same mistake until she and I saw and filmed one here" at Wagner.

Acorn's story reminded me of something that Ben Rostron had once told me about Wagner's orchids. On a good day, the average visitor will see maybe four or five species. But if one knows where to look and how to look, they will find more. "Intimacy," as bryologist Robin Will Kimmerer once wrote, "gives us a different way of seeing when visual acuity is not enough."[11]

The thing to remember about discovery, I have come to realize after spending so much time with entomologists and peatland ecologists, is that it usually doesn't happen. Most of the expeditions and surveys end up with no new significant insights and no new or rare species found. On a really bad day, Hawaiian botanists like Steve Perlman and Ken Wood bear witness to a scene in which "Extinction" is the title of the last chapter.

The fact that we don't know so much about what is going on in peatlands underscores why we need to establish benchmarks for insects to better measure the success of peatland restoration and how well virgin peatlands are responding to climate change.

This, however, was not a day for disappointment. As Acorn and I hiked along the northern edges of the fen, we ran into Kevin Judge, an entomologist based at MacEwan University in Edmonton. He was looking for Roesel's katydid, a grasshopper-like insect which had never been seen in this part of the world until he found one here in 2017. It was a head-scratcher at the time, because the nearest known habitat was more than 900 miles away. It continued to be elusive until Judge found one later that day.

My last impression of Wagner was one that presented itself as Acorn and I were walking along the side of the four-lane highway, heading to the small, sandy parking lot. The sound of cars and trucks passing by was so loud that we gave up trying to communicate. It was there, along a culvert by the side of the road, that I saw not one or two lady's slippers, but three beauties. They were so close to the road that cars and trucks were likely slapping them with water on rainy days.

What I could not figure out was what they were up to. Were they fragile survivors hanging on desperately as road and industrial developments closed in on them? Or were they the resilient ones, tough enough to withstand the abuse and spread out into new habitat? I took a photo, recorded the exact location, and promised myself that I won't come back in the coming years to see if they were still there.

Chapter 7

Peat and Reptiles

The bog floor shakes
Water cheeps and lisps
As I walk down
Rushes and heather.
— Seamus Heaney, "Kinship"[1]

Staring down into the water, where light from the late afternoon sun refracted into a kaleidoscope of colors, the one thing that stood out on my first canoe trip on Georgian Bay's French River was the dark silhouette of a white pine parallaxing in ghostly fashion. The pines, smooth rock, and dark water evoked an intimidating sublimity that did not reveal its secrets readily. If you were lucky back then, as I was sitting by a campfire on a starry night, you could hear the howl of a wolf, or maybe, as my companion swore, see its shadow lurking in the damp forest beyond. But you didn't know what else was out there, because venturing into the bogs and fens beyond the riverside was not the thing to do at night.

Georgian Bay is situated in a corner of Lake Huron where there are 30,000 islands of pinkish-gray granite. Many are dotted with the cottages of affluent New Yorkers and Torontonians who can do the drive in a day or two. It's the world's biggest freshwater archipelago. The bay is famous for those wind-swept pine trees that magically grow out of slim, peat-filled frost fractures in the Canadian Shield.

It's a Group of Seven art scene that never gets tired because it is still, culturally speaking, a relatively fresh Canadian landscape image. Up until the time painters A. Y. Jackson, Lawren Harris, Arthur Lismer, and their fellow Group of Seven artists, as well as Tom Thompson (who died before the formation of the group), began venturing north into this part of the world in the 1910s and 1920s, sketchers and painters made Canada look like a country dominated by hills and dales filled with butternut, maple, oak, and lakes that resembled those in the English and European countrysides. It was a metaphorical way of taming the chaos of colors and the otherworldly tangle of a boggy landscape. "Healthy, lusty colour which you see in Canada is no doubt considered vulgar," was the way artist J. W. Morrice described it in 1910 when he complained to a friend that English art dealers were "poisoning everything" with their " ghostly Dutch monochromes."[2]

More than anyone else, members of the Group of Seven informed Canadians that their "North" was unforgivingly boreal—spongy, buggy, running with wolves and bears, swimming with beavers and snapping with turtles more than it was pastorally Carolinian and populated with noble stags leaping over babbling brooks. "We the North," the rallying cry for the Toronto Raptors NBA basketball team, was one of the many identity-shaping truths that was firmly rooted in the culture that grew out of the Group of Seven paintings.

Many Canadians at the time resisted this reality. Hector Charlesworth, a Toronto-based art critic, was appalled when he took in a Group of Seven exhibit in Toronto in 1921. In reviewing Lawren Harris's painting *Beaver Swamp*, Charlesworth wrote: "A while ago I saw in an art gallery, a . . . picture . . . of a swamp. A repulsive, forbidding thing. One felt like taking a dose of quinine every time one looked at it. If ugliness is real beauty they have yet to prove it to a very large mass of the assembled public." Charlesworth much preferred artists like Homer Watson, who "did not paint the wilds, but the hills and valleys that the pioneers of Upper Canada made opulent and fruitful."[3]

Indigenous people like the Anishinaabe of the Great Lakes region and the Cree of the seven First Nations (Mushkegowuk) living northwest of them in the Hudson Bay Lowlands, the second largest peatland in the world, saw it differently. The English word *muskeg* comes from the Cree word *maskek* and the Ojibwe word *mashkiig*, meaning "grassy bog."

Their insights into muskeg never counted unless they came from someone like Grey Owl, a binge-drinking Englishman named Archie Belaney who had five wives while successfully passing himself off as half-Indian in his public appearances and on the back covers of his internationally best-selling books on nature. Grey Owl did live as a part-time trapper in the boreal forest, and he knew a thing or two about the natural world, thanks, in part, to Angele Egwuna, an Anishinaabe woman who was his first and only legal wife. But his descriptions of muskeg are quintessentially British—"unpenetrable," "dirty," "impassable" places to be avoided. His beavers are presented more as pets than as wild animals that avoid humans. (For some time after his death in 1938, people ranked him with John Muir and Rachel Carson in the pantheon of environmentalists. Richard Attenborough flattered him in a film he made about his life.)

There was little appreciation back then that the Anishinaabe and other indigenous people used peat for diapers and sanitary napkins and harvested food that grew out of it. No one knew or cared either that they harvested medicinal plants like Labrador tea in bogs and fens. It wasn't just a handful of plants that were on their list of pharmaceuticals. When forest ecologist Hugo Asselin and his colleagues tried to determine how many kinds of medicinal plants there were in the boreal forest that indigenous people were homing in on, they came up with a list of 546 plants that were used to treat 28 diseases.[4]

I never equated Georgian Bay with bogs and fens, or with swamps and marshes, until I completed that first canoe trip and then looked beyond the pine, granite, and lakeshore vistas of many of the Group of Seven scenes and recognized the splashy shapes and wild licks of color that belong to those wetlands. They are not the most famous of Group of Seven paintings, but they are the most interesting because they brought an awful splendor to a landscape that requires so many words, often contradictory, to describe it.

On this latest visit to Georgian Bay, it was the rattlers in these waterlogged peatlands that intrigued me the most. I had always associated them with the Mohave Desert, where I was earlier in the year, and the arid badlands of the West close to where I live.

It was one of those warm, humid days when a cool gust of wind off the lake would have been welcome. Alana Smolarz was snake hunting in a bog. She was wearing long pants, knee-high boots, a sweater over another shirt, and a tan-colored balaclava to keep the blackflies from biting. The dark sunglasses made her look a little like a comic-book bank robber, but not nearly as odd-looking as two bog researchers I had met earlier in the day. They were wearing hip waders and snowshoes so that they would not sink into the peat.

Cory Kozmik, an environmental management biologist for Magnetawan First Nation, was walking along behind us dressed in similar, but slightly more fashionable clothing.

Smolarz was a species-at-risk biologist at Magnetawan First Nation. She was leading me along in such a cheerful way that I stopped at one point and asked if she wasn't worried about stepping on a snake and getting bitten. A rattler, I was convinced, would see us before we spotted it in this mire of moss, sedge, giant reed grass, and carnivorous pitcher plants.

"No, not really," she told me as she continued to sweep aside the vegetation with a metal rod. "The venom of the massasauga rattler is pretty mild. And they're not aggressive unless you poke them and get them riled up. They give you advanced warning with their rattle."

"Finding them, though, is a little tricky," she allowed. "It's a little like searching for Waldo. You know there's at least one out there, but they blend in so well that's it's hard to know where to look."

Smolarz studied with Mike Waddington, an ecohydrologist who began his career in the peatlands of the Hudson Bay Lowlands nearly thirty years ago before branching off in so many directions that it was difficult for me to track and pin him down. With luck and a little persistence, I convinced him to take me on a tour of several peatland study sites just north of the Muskoka region of Georgian Bay. It was he who introduced me to Smolarz.

Alana Smolarz made a name for herself as a graduate student when she established a startling connection between rattlesnakes and peat. With the help of radio telemetry that allowed her to track the movements of the rattlers, she discovered that they were hibernating in the peat beds of raised bogs, preferably at depths where they wouldn't

The carnivorous pitcher plant that grows in nutrient-poor bogs such as Algonquin Provincial Park has the ability to trap and consume everything from insects to juvenile salamanders. (Photo: Edward Struzik.)

drown or die of hypothermia in rising groundwater or freeze in frost layers creeping down from above.

The fact that no one knew that rattlers nest in peat was yet another reminder of how little is known about peatland ecosystems, and how they support so many endangered and threatened species in so many different ways.

When considered carefully, a sphagnum peat bed is the best place to be if you are a reptile living this far north, because sphagnum absorbs 16–25 times its weight in moisture. It has an R-value that once made it a popular form of insulation in old homes like mine until homeowners like me learned that the mass of peat shrinks precipitously as it dries out.

Smolarz, Kozmik, and their colleagues have caught more than a hundred rattlers. Many were tagged and tracked to figure out where they mate and hunt in spring and summer and where they hibernate for half the year. Some snakes, like a much-admired gestating female they caught, are given names. Fefe is what they called her. The Don is a monster male I learned about an hour earlier, when Kozmik showed me where she found him sunning himself on a granite rock along the edge of a bog.

"The Don," she told me, came to mind when she saw that he had ten rattle segments. (The rattle is composed of a series of hollow, interlocked segments made of keratin.) His rattle was so large, it purred. It's why she gave him a boss-like name.

It's not just the snakes that have an affinity for peat. Turtles are also on the list of creatures that Smolarz and Kozmik are tracking. They too get names such as Ronaldo, Xavier, Littlefoot, and Darwin.

The biggest threat to snakes and turtles in Georgian Bay are roads, mines, poachers, logging, and cottage developments that kill, steal, or displace them. When cars and trucks come and go, reptiles are just too slow to get out of the way.

I got a lesson in that when I arrived the day before and spent the time with Mike Waddington, visiting a couple of other study sites near the town of Nobel, which was named after Alfred Nobel, the inventor of dynamite. Two First World War munitions factories were built at a site surrounded by swampy peatland, presumably to keep people away.

Each year, Waddington and his students at McMaster University give out the "Nobel Prize for Peat" to the scientists who produce the best scientific paper on the subject.

On the drive to Nobel, Waddington pointed to a black dot on the road that looked like a clump of turf that had fallen off the back of a muddy truck. Like most drivers coming from the city, I was unaware that it was a turtle, let alone a threatened Blanding's, which is a medium-sized turtle that distinguishes itself from other turtles with a dome that looks like an army helmet. Its bright yellow chin and neck make its brown eyes look even bigger than they are.

Blanding's turtles normally spend their time in the shallow water of bogs and marshes, unless they are searching for a mate or looking for a place to nest. That's when they slowly cross roads like the one we were on. Snakes get killed as well. It got pretty ugly at one point. In the early 2000s, three students from the University of Guelph spent a couple of summers counting dead reptiles on a secondary road in the area. They found 296 snakes and 71 turtles. Forty-four of them were species at risk—reptiles like the Blanding's as well as the spotted turtle and stinkpot turtle. Stinkpots, I learned, give off a musky, skunky odor when they are disturbed. Like the critically endangered Indochinese box turtle, they also climb trees.

When the Group of Seven ventured into the peatland of Georgian Bay, much of the landscape was already "slashed up, burned over, and flooded," as A. Y. Jackson described it. Elk, as well as the white pine, were almost all gone. When the loggers left, tourists came to claim a piece of this summer paradise. Most of them were unaware that their annual drives north and south were destroying the swamps that were filtering their water, protecting their cottages from wildfire, and nurturing the turtles, snakes, and howling wolves that reminded them of that much-coveted communion with nature.

The Magnetawan First Nation is now working with scientists and the Georgian Bay Biosphere Reserve to put a stop to the carnage on the roadways and other developments that are undermining the integrity of the wetland habitat. To them, rattlers are not slithery, venomous reptiles that need to be feared and avoided, but guardians of the berries, root vegetables, and medicinal plants that are harvested annually. They are one of the many "other than human" animals in indigenous cultures that offer wisdom and protection, kinship, and ceremonial symbolism.

The alliance among the Magnetawan First Nation, biologists, and local residents behind the Georgian Bay Biosphere Program has chalked up a number of success stories over the past twenty years.

One of them involved Chantel Markle, a postgraduate student in Mike Waddington's lab. She also made a name for herself with a study that showed that a properly designed, 3.6-kilometer-long fence that was built along a highway connecting Long Point Peninsula on Lake Erie to mainland Ontario was extremely successful in preventing turtles and snakes from being run over. She and her colleagues found that similarly designed fences and culverts that have been built along Georgian Bay's roads are allowing some turtles and snakes to safely get from one side to other.

"There is really nothing we can do about the roads that we have already built, so fencing is the best way of preventing turtles, snakes, and other animals from getting run over," said Waddington as we passed the turtle just as it made it off the road. "What we can do in the future is to make sure that we don't build roads that will run through critical habitat. That would be the ideal way of preserving biodiversity."

Ecohydrologists who study peatlands are a rare breed, as I have found on my journeys to the Great Dismal Swamp, the Pocosin, and other peatlands. Their strengths lie in an unlikely combination of physics, engineering, hydrology, soil science, botany, and wildlife biology. In addition to the biting flies, that may be why there are not a lot of them.

If anyone can handle blackflies, it's Waddington. He spent many of his childhood days in Georgian Bay country kayaking, canoeing, and picking blueberries with his mother and father, Jim and Sue, who have spent nearly forty years locating and photographing the landscapes in paintings by the Group of Seven. (Small as the world can sometimes be, I shared the stage with them at the 2016 Wilderness and Canoe Symposium in Toronto.)

Waddington has another skill that helps him navigate through the labyrinth of patterned bogs in which the geometric strings and flarks (depressions) that are associated with them would confound even the most gifted backcountry enthusiast looking for a way in or out. Not so long ago, Mike was a world-class orienteer, excelling in a competitive sport that involves running, walking, and sometimes climbing from one designated point to another. The difference between orienteering and backcountry racing is that in orienteering there is no prescribed route between Point A and Point B. The winner is the person who independently navigates the course—not necessarily with the shortest route, but with the quickest time. To get from one point to another, competitors can use maps and a compass, but not GPS.

Waddington was introduced to the sport when he accompanied his father, a physics professor, on a sabbatical to Norway, where orienteering is almost as popular as cross-country skiing, a national obsession in winter. He got hooked while he completed his PhD in Sweden, fifteen years later in 1993. It was a time when the draining of peatlands was a means of driving the Scandinavian economies with fuel derived from peat.

Like the Irish, Scots, Estonians, and Russians, the Scandinavians still harvest peat for energy, though not nearly as much as they did in the past. They also drained peatlands to grow trees for commercial purposes.

When Waddington was in Sweden on his own sabbatical in 2006, scientists there were busy trying to figure how much peat they had removed from the landscape—95 billion cubic meters, according to one estimate—and how much they had left that could be practically harvested. They figured they had about a 2,500-year supply, which did not include peat in protected areas.[5] Few people back then were concerned that 300,000 hectares (nearly a quarter million acres) that were drained for plantations did not produce the trees that were expected.[6] There was just so much peat that it didn't matter.

The degradation often happens incrementally and so slowly that most people don't notice when a peatland is no longer effectively filtering water, mitigating flooding, or storing water when it is needed most. The Wainfleet Bog that Waddington is also studying is one of the few acidic bogs left in southern Ontario. It was once three times the size of Manhattan. Only 6 percent of it remains. Sifton Bog in London, Ontario, is one of the few that survived after flood-control dams were erected on the Thames River. The Fanshawe Dam in London reduced flooding, but the degradation of the peaty bogs that would have otherwise absorbed excessive moisture left the city vulnerable in 2000, 2008, and 2009, when floods overwhelmed the dam. And as the river valley dried out the bogs, invasive species moved in, outcompeting many native plants.

"Peatland ecosystems are pretty resilient," said Waddington as we hiked into a soggy bog where every step came with a squish. "They can withstand drought, wildfires, and roads. But they are very vulnerable when two or more of those things, such as a road, a drought, and a wildfire come, into play in a short period of time."

Up until recently, an intense wildfire seemed like an unlikely possibility in the Muskoka region because it is so wet and humid. But cottage owners got a wake-up call in the summer of 2018 when a fire burned 11,000 hectares (27,000 acres) of forest to the northwest of the region, compelling wildfire fighters to bring in water bombers. The fire could be written off as an anomaly if it weren't for the fact that wetlands like the Pocosin and the Everglades in the south, as well as boreal bogs and fens in the north, have been burning bigger and more often than at any time in recorded history. Even Sweden, a country that knows almost nothing about big peatland forest fires, experienced a burn in 2018 that was so intense that help had to be brought in from Germany, France, and Italy.

"Wildfire fighters used to be able to rely on water-logged peat bogs and fens to help slow or stop a fire in its tracks," Waddington told me as we nimbly walked across a narrow boardwalk that took us to the edge of an algal pond. "But increasingly, we're seeing fires like the ones in Sweden and the ones that forced the evacuation of Slave Lake in 2011 and Fort McMurray in 2016 burn through peatlands that have dried out."

The Horse River Fire in Fort McMurray was the costliest natural disaster in Canadian history. Nearly 90,000 people in the oil sands capital of the world were forced to flee their homes at the last minute. Veteran forest-fire fighters were caught off guard by the speed at which the fire moved because their vegetation maps didn't account for the fact that many of the bogs, fens, and marshes that were in the path of the inferno had either dried out or been overtaken by highly combustible black spruce trees. Waddington was one of several scientists who were involved in a postmortem of sorts. He and his colleagues assessed a peatland about seven miles south of Fort McMurray that had been partially drained for a forestry experiment. What they found is that the fire that burned the drained area was much more extreme than expected. Instead of slowing the fire, it helped it gain momentum.

"The basic lesson we learned is that a healthy wet peatland is a firefighter's best friend," said Waddington. "A dry or degraded peatland is the enemy. Once the peat begins to burn it is hard to put it out because it burns downward into the ground, where no amount of water bombing can put it out easily."

No single peat fire can be attributed to human-caused changes to the climate and the landscape. But since at least 2007, there have been a growing number of head-spinning peatland fires in the northern hemisphere that point to a pattern of fiery events that are impossible to attribute to natural cycles.

Scientists like Mike Waddington believe that climate change will result in an increase in peatland fire severity and frequency. Not only will this liberate enormous amounts of carbon and methane that have been frozen in the soil, but it will alter the minerals in the soil and nutrient supplies in ways that favor woody shrubs over moss, sedges, grass, and lichen.

When I asked Waddington what the solution might be, his answer was deceptively simple: "We need to selectively remove some of the black spruce trees that have taken over those peatlands that we've drained or degraded, and rewet peatlands that we have allowed to dry out. We can grow super-mosses that are better than any kind of fire retardant out there. It's not cheap. But we've done it here in Canada. In the long run, it will be cheaper than it costs us to fight fires and deal with expenses associated with poor air quality and a long list of things that will need to be done to mitigate the impacts of climate change."

Peatlands have a set of values that have been both appreciated and denigrated for centuries. One generation sharing these virtues and vices with the generations that follow leads to a meaning that is as cultural as it is scientific. British ecologist Oliver Rackham once described four ways in which a landscape is lost.[7]

The loss of wildlife such as rattlesnakes and turtles is one of them. The loss of beauty like the mists on a bog that are slowly burned off by warm shafts of sunlight is another. The loss of freedom speaks to the connectivity that allows peatland creatures like the caribou and turtle to move in and out of fens and bogs in order to escape wildfire or to find a mate. The fourth is loss of meaning, which would happen to members of the Magnetawan First Nation if the rattlesnake, the protector, disappeared.

The Native American story about bogs that speaks to the loss of meaning is one that Fred Pine, an Anishinaabe elder from Garden River First Nations near Sault Ste. Marie, Michigan, shared with Michael Wassegijig Price, an educator who believes in the importance of integrating science with indigenous knowledge.

In the story "Genondahway'anung," a rising star with a flaming tail is about to crash into the Earth. The Great Spirit, Gitche Manitou, warns the Anishinaabe and tells them to take refuge in a bog and roll themselves up in the moss and mud to protect themselves. The star crashes. The forest burns. Only those who rolled themselves up in moss survive.

Price suggests that the story may have originated in 1871, when the Peshtigo Fire tore through the peatland forests of northern Wisconsin, killing as many as 1,200 people. (Some people at the time believed a meteorite caused the blaze.) We know why the forests burned. So much land had been drained, and so many trees had been cut down for agricultural purposes, that once the slash piles were set on fire on a hot windy day, the inferno took off like a freight train. What we don't know is what stopped it from doing more damage and harm than it did. I suspect it was the peat-accumulating wetlands that dominated much of the landscape back then. Once the fires reached their spongy borders, there was nothing dry enough to keep them going.

Chapter 8

Mountain Peat

To love a swamp, however, is to love what is muted and marginal, what exists in the shadows, what shoulders its way out of mud and scurries along the damp edges of what is most commonly praised. And sometimes its invisibility is a blessing.

— Barbara Hurd, *Stirring the Mud: On Swamps, Bogs, and Human Imagination*[1]

Fran Gilmar still looks like the nineteen-year-old woman who moved to Crowsnest Pass along the Montana–Alberta border 58 years ago to teach figure skating to the children of mountain coal miners. She is thin and graceful for her age, and she continues to pull her hair back in a ponytail. She offered a bright smile, along with coffee and donuts, when we met along a gravel road at the base of Grassy Mountain.

Those coquettish looks and twinkling eyes were deceiving, as I found out when two men on a quad and motocross drove up on that dusty gravel road that runs along her remote mountainside ranch.

Gilmar, widowed since 2003, is one of the few people who have access to this heavily gated mountain wilderness. She was hunched over, locking one of those gates, when we heard the men coming. Instead of turning the other way or offering a polite "What are you doing on my land?" she stopped the pair in the middle of the road and ordered them to "Get the hell out. You're trespassing." She has no use for off-roaders tearing through the wilderness, as they often do. When one of the men objected, she cut him off and repeated what she had said—only louder, more emphatically, and with a frantic shooing of both hands.

I wasn't sure how this standoff was going to play out. There are exceptions, but off-roaders have a reputation for being belligerent. It's been a long-standing issue in the Crowsnest, and it continues to infuriate residents, ranchers, and backcountry enthusiasts who come for solitude. Both men must have realized that, win or lose, taking on a 77-year-old woman in front of witnesses was not going to end in their favor. In relatively short order, they turned around and never looked back.

It was no surprise to see Gilmar so hot under the collar. She had just taken me and a friend on day-long tour of the area, which a subsidiary of a giant Australian mining company plans to dig up over the next quarter century in order to ship coal to countries such as China and India. At the time of my visit, Gilmar was one of a small number of landowners refusing to sell. She agreed to show me why when local wilderness guide Heather Davis reached out on my behalf.

"This is not for me, this is for the water, and the land that pays homage to the water flowing out of this mountain," Gilmar said after the off-roaders departed. "It's for cold, clear mountain streams like Morin and Green, and the fen on my land that flows into Gold Creek. It's for the westslope cutthroat trout, which are almost gone from this world. It's for this water that flows through ranches like the one I bought with my inheritance. It's for the water that spills into the Oldman River, which flows in the South Saskatchewan and through the prairies."

"Oh no! I'm not going to sell out to those people," she said looking at me sideways, nodding her head from side to side. "Water is already scarce in these parts. I'm going to be a burr up the ass of this company for as long as it takes to stop them."

It was mid-October, when the short days in the foothills are still warm but the nights are long and cool enough to turn the leaves of balsam poplar and aspen into gold coins. This used to be white pine and limber pine country—gnarly, shallow-rooted trees that are well enough anchored in thin, rocky soils to withstand the super-charged chinook winds that periodically blast through. "Snow eater" is what the Blackfoot people call these blows. These pines, however, are not long for this part of the world. Most of the pines in the Crowsnest are ghost trees that serve more as roadside attractions or markers for backcountry hunters than for the food their seeds once were for resident birds like Clark's nutcracker. The beetles, blister rust, and climate change are killing them.

I cast my first fly in these mountains years ago, when a friend took me to his favorite spot along the upper reaches of the Oldman River. He promised that even a novice like me would have a good chance of catching a rainbow or bull trout and maybe a cutthroat. He was right. Seeing the fringe of green moss in and along the mountain stream that runs alongside Gilmar's ranch, I wondered why I had never gone back to that spot. The fishing was that good.

Gilmar talked as if she were in a pulpit, driving the devil from her congregation of moose, elk, wolves, bears, cougars, trout, and the rare plants she covets. She preached in this manner all day long while she showed us sites where rare plants grow, streams in which westslope cutthroat trout spawn, and the mountain fen bordering her property. Typically, after each stop, she would walk back to her rusty pickup truck to take us to the next hidden treasure. And then, almost without fail, she would stop after a minute of driving, walk back to our truck, motion for us to roll down the window so she could add a few more details to the sermon she had just ended. After a day of this, I was amazed that she hadn't collapsed.

Benga Mining Limited, the wholly owned subsidiary of Australia's Riversdale Resources Limited, was one of four companies actively exploring six potential mine sites at Grassy Mountain. Other companies have been boring holes along the mountainsides to the northwest.

There are those like Gilmar who oppose coal mining for the dust it will produce and for the damage it will do to streams, meadows, and fens, and those who support the mining companies for the jobs they promise to bring to a region where shuttered storefronts are a common sight. Coal has broken the hearts and lives of miners off and on here since 1903, when 82 million tonnes of limestone came tumbling down Turtle Mountain, rolling over the entrance to one of the region's first coal mines. More than seventy people died.

Coal lived on, but with only the littlest to offer in wages. After a bitter eight-month-long strike in 1933, the impoverished citizens of Blairmore used an "X " on their ballots to express their anger. They elected Canada's first Communist mayor, town council, and school board. For a time, one of Blairmore's streets was named after Tim Buck, a Communist leader and labor organizer who spent two years in jail from 1932 to 1934 for sedition.

The labor disputes rarely ended in the miners' favor. One by one, the coal mines shut down. The last one was shuttered in 1983.

Without an industrial tax base, the local government council has been struggling to provide services to its residents, including a growing number of retirees who are moving to Crowsnest Pass to take advantage of cheap real estate in a gorgeous mountain setting. Many of the old and new residents had hoped that the Pass would become as prosperous as the tourist-magnet Rocky Mountain town of Canmore near Banff has since its last coal mine closed in 1979. Most everyone knows that won't happen if the Pass returns to its coal mining roots.

Crowsnest Pass is sandwiched on all sides by wildlands such as the Castle to the south, Bob Creek to the east, the Beehive to the west, and Waterton Lakes and Glacier National Parks to the southeast. The Crowsnest, Livingstone, Oldman, Castle, and Waterton—all wonderful trout-fishing streams—flow through them with cool, crystalline clarity even after periodic wildfires, logging, and warming temperatures that should have heated them to levels lethal to fish. It's the trees, the spongy mosses, and the frigid groundwater that continue to keep many of these mountain streams cool, according to Uldis Silins, a forest hydrologist I accompanied on two occasions.

Coal was not supposed to be part of the picture after the last mine closed. But that all changed when the Alberta government, short on the revenue that once came reliably from the oil sands industry, signaled the end of its decades-long coal policy, which banned open-pit mines in many parts of the Rockies. The signal, however, was sent to the coal-mining companies, not to ranchers and indigenous people.

In anticipation of what they apparently knew was coming, mining companies started wooing local politicians and business people, going so far as to pay for the cost of relocating golf-course fairways and a golf and country clubhouse so that a coal-loading facility and a rail loop could be built in Blairmore. It was then that ranchers began to realize that something was up. Only when they started asking questions was it made apparent the government had opened up 1.5 million hectares (3.7 million acres) along the east slopes of the Rockies to coal mining.

When I first learned of plans to lop off these peaks, I wondered how many ephemeral mountain streams, how much groundwater, and how many fens and wetlands might be impacted. It was impossible to know, because many have still not been mapped. Looking through Benga's environmental impact statement for Grassy Mountain, it didn't look like much in their case—just 17.6 hectares (43.5 acres) of fens. That was all that the company had to say. There was no mention of rare plants or animals, the source of groundwater, and how the groundwater is interconnected to other water bodies in the region.

I wasn't surprised by the omissions. I have hiked along many mountain peatlands without recognizing them for what they are and the role they play in an alpine ecosystem. The peatlands around Helen Lake in Banff and Amethyst Lake in the Tonquin Valley of Jasper National Park (possibly the prettiest place in the Canadian Rockies) are some of them.

I wondered whether there was more to it here in Crowsnest Pass. Gilmar assured me that there was when she took us to that mountain fen that borders her property.

"This place is not for people," she said, clinging to willows to steady herself as we descended into the fen. "We had a horse sink down to its knees here before she figured out that the best way of saving herself was to fall back on her big behind so she could stabilize and pull herself out. We almost lost the girl who was riding her."

To an uneducated eye, mountain fens can be hard to identify, because they often lack surface water during the late-summer and fall months. It's not until you start squishing through them that you appreciate how the perennial flow of groundwater saturates the ground. I learned this the hard way several years ago when I took what I thought was a shortcut through a soggy alpine peatland in Kluane National Park in the Yukon, where I once worked. It knocked two miles off my journey, but added an hour of grief getting through the boot-sucking mire.

Groundwater from her fen, Gilmar told me, drains into Gold Creek, where the westslope cutthroat trout spawn. The fen has no name, she allowed, because no one other than she, her children, and her late husband knows that it exists. It's too far off the beaten track.

Gilmar looked down just then to examine the muddy, indistinct track of an ungulate that had recently passed through. "It better not be a moose," she said. "The eyes of a bull at this time of year are rolling back into their sockets, and what little it sees is red. You're done if there is no place to run to."

Moments later, the sound of something nearby caused Gilmar to pause and point her index finger up into the air. "There's no fooling around with that bear," she said. "She has a cub to protect. She's down there somewhere eating berries."

I had once assumed that fens and bogs were predominately lowland ecosystems, sustained by precipitation and groundwater flowing from higher elevations. But as mosses and other peatland plants revealed themselves in small, high-elevation wetlands like those I saw on hikes and climbs in Banff, Kootenay, Yoho, Kananaskis, the Sierra Nevada, the Beartooth Mountains of Wyoming, the San Juan Mountains of Colorado, and in the Alaka'i Swamp of Hawaii, I began to see things differently.

In Kentucky, a rare Appalachian seep/bog is perched on the crest of Pine Mountain in the E. Lucy Braun State Nature Preserve, named after one of the twentieth century's most influential botanists. Mountain fens and bogs like this in the Appalachians are among the rarest and most imperiled habitats in the United States. They are home to mountain sweet and purple pitcher plants, tiny bog turtles, poorly understood upland burrowing crayfish, rare Appalachian brown butterflies, and three-inch-long gray petaltail dragonflies that do well where sphagnum and club moss grows.

Reaching a height of 3,700 meters (more than 12,000 feet) in the Tibetan highlands of China, the Zoigê Marsh is the most extensive mountain peatland in the world. Fens that cover much of the 10,000-square-mile plateau have peat that is up to thirty feet deep.

Most mountain peatlands are fens, typically small ones. They cover 1 percent of the land surface in the Beartooth Mountains of Wyoming, and 1 percent of the San Juan Mountains in Colorado.[2,3] Their distribution is worldwide. Two and a half percent of the Snowy Mountains in New South Wales, Australia has peat.[4] Even the autonomous region of Galicia in northwest Spain has peat accumulating in its mountains.[5]

In each case, the distinct nature of these mountain fens plays an oversized role in supporting insects, plants, and animals and in storing water and carbon. Small peatlands such as the eighteen that were inventoried in Wyoming contain 32 threatened plant species. Four of them—the small round-leaved orchid, bearberries, and the low blueberry willow—are found nowhere else in the state.

David Cooper is a Colorado State University scientist who has done more for the understanding of mountain fens than any other scientist. His research has taken him to the high country of the southern Rockies, the Sierra Nevada, the Cascades, the Pacific Coast Range, the Carpathian Mountains of Poland and Slovakia, and the Andes near Cajamarca, Peru, where indigenous llama, alpaca, and sheep ranchers graze their animals on *bofedales*, the word they use to describe peatlands that are formed and sustained by groundwater and meltwater from glaciers.

The High Creek Fen, situated nearly 10,000 feet up in the Colorado Rockies, is one of his many marvelous discoveries. American explorer Zebulon Pike probably took no serious notice of this wetland when he and his party passed through in 1805. Kit Carson may have paid more attention when he came in to trap beavers. I almost missed it when I stopped by en route from a hike I had done along the Grand Ditch Trail to Pike's Peak, one of the few places in North America where you can hike to 14,000 feet in a day.

A small stand of blue spruce caught Cooper's attention when he passed through the area early in the summer of 1988. The tree is not common in the valley. When Cooper hiked in to get a closer look, he realized that the clump of trees was growing out of a sedge-covered fen sustained by water trickling in from High Creek, and especially from the groundwater constantly flowing from the limestone and dolomite glacial outwash and karst pools that connect to the underlying red beds (sedimentary rock colored by ferric oxides). Cooper knew he had stumbled upon something special. He didn't fully appreciate how special, though, until he invited William Weber, the curator of the University of Colorado Herbarium, to have a look later that summer.

Mountain fen in the Sierra Nevada, Sequoia National Park. Peatland ecologist David Cooper has done more for the discovery and understanding of mountain fens than anyone else. (Photo: David Cooper.)

Weber and others identified fourteen rare plants. There are only a very few extremely rich fens like this in the United States, which is why The Nature Conservancy eventually purchased the land.

Since then, Cooper, along with colleagues and students, has identified thousands of high-elevation fens that were previously unknown or unappreciated for what they are. The numbers, in some cases, are mind-boggling. In 2008, 1,738 fens covering 11,034 acres were identified in the Grand Mesa, Uncompahgre, and Gunnison National Forests of Colorado. Ninety percent of these fens were found at elevations ranging from 9,000 to 12,000 feet. Cooper estimates that in the San Juan Mountains alone there are more than 2,000 fens, ranging in size from 0.5–50.7 acres (0.2–20.5 hectares).

In Colorado, there are rich fens and poor fens as well as a very small number of iron fens, which are extremely rare and acidic due to the oxidation that occurs in the pyrite-rich bedrock. What all of them have in common is a diversity of rare plants and insects that far exceed the fens' small size. And each one stores carbon just as well as peatlands in western Canada are noted for doing.

Cooper refers to these high-elevation fens as "little pieces of another world." The pollen and plant materials are so well preserved that they offer a glimpse of the mountain landscapes as they were 10,000 years ago, when the ice sheet began to retreat.

Many of these fens have been degraded by mining, agricultural development, and water-diversion projects like the Grand Ditch, a canal that was cut into the slopes of the Never Summer Range between the 1890s and 1930s to supply irrigation water to farmers along Colorado's northern Piedmont. The canal continues to divert a significant amount of water away from the wetlands of the Kawuneeche Valley in Rocky Mountain National Park.

For a long time, beavers in the valley mitigated the damage by building dams that held back enough water to sustain the peaty meadows. But around the turn of the century, the beavers moved out when the aspens and willows they used to build their dams mysteriously began to disappear. Initially the blame was put only on moose and elk that were moving into the region. Moose and elk browse on willows. But then Cooper and his students realized that sapsuckers might also be playing a role by chipping into the bark of the willows and allowing a deadly fungus to enter. Those stems did resprout. But they did not last long because there are just too many moose and elk eating them up.

Walking along that fen at Gilmar's ranch, I wondered whether there might be an elusive dragonfly or moth, or maybe an endangered orchid or some other rare plant growing there. Was this place important to the grizzlies that had only recently returned after being absent for decades? If a wildfire burned through this hot, dry corner of the Rockies, as wildfires often do in this part of the world, would it be slowed or stopped by this fen? If so, would it open up the landscape, keeping sphagnum and other peatlands in place and preventing them from being taken over by ground cover that does not store carbon as well? I considered the likely possibility that there are many more fens that are unaccounted for in the Canadian Rockies.

Mac Blades offered some clues when I dropped by his ranch north of Grassy Mountain, on the rolling plains that stretch out from the foothills to the breathtaking Williams Coulee. I had first met Blades, a number of years earlier, when I convinced him to allow me to ride along on the annual cattle drive that he and his extended family do each year to get their cattle in and out of a grazing lease along the mountainsides of the Oldman and Livingstone River watersheds.

It was a memorable trip, not just for the scenery and the verdant meadows we passed through, but also for the fact that I had never ridden a horse before. Stories I was told about the first cowboys there shattered the image I had had of Canadian cowhands being dull compared to their southern counterparts. Most interesting was John Ware, an African American who was born into slavery in 1845 on a cotton plantation in South Carolina. When the Civil War ended, Ware used his newfound freedom to get as far away as possible from the former slave state. He did that by going to Texas and getting hired on cattle drives to Montana and then from Montana northward to the Bar U Ranch, which is northwest of where Blades grazes his cattle.

Ware worked at the Bar U right up until the time the Sundance Kid was hired on as a dollar-a-day horsebreaker. Whether they met, no one knows. What we do know is that through hard work, Ware created a family and a fine ranch of his own.

It would have been a wonderful story had he and his wife, Mildred Lewis, not died tragically in different ways the same year, leaving six grief-stricken children to live with their grandparents in Blairmore in Crowsnest Pass, near the base of Grassy Mountain. Mildred died of pneumonia shortly before Ware was killed when his horse tripped in a badger hole and collapsed on top of him. His funeral in Calgary was the biggest ever up until that time. Ware Creek, John Ware Ridge, Mount Ware, and Ware Creek Provincial Recreation Area—places I have hiked—all bear his name, as do a school and government building.

Mac's grandfather knew John Ware. His mother and aunt were good friends with the daughters and knew the three sons. Mac also knew some of their children and remembers them fondly.

It's hard to put into words how pristine this part of the world has remained since Blades' grandfather was drawn to the area in the

1880s. Clint Eastwood may have expressed it best by choosing to film *Unforgiven* in these parts because it was the only place his crew could find outside a national park where there were no fences obstructing the views.

Well into his seventies, Blades looked as youthful as he did on that cattle drive more than a decade earlier. When I told him how impressed I was that he still sitting tall in the saddle, he smiled and thanked me for the compliment before removing his cowboy hat to show me a head of gray hair. He then invited me to come to the dining room table, where he pulled out some maps that showed how the coal-mine exploration is playing out in the mountain meadows where he herds his cattle.

"The mountain that is slated for mining is the one between the Livingstone River and the Oldman River," he said pointing on the map to an area that I was familiar with. "These rivers are the headwaters of the Oldman River system. This is the main source of water for the drainage."

By the time of my arrival, 420,000 hectares (over a million acres) of government land in southwestern Alberta had been leased to coal companies. Blades and his fellow ranchers are justifiably concerned that high levels of selenium that are likely to be released by the coal mines will pollute these watersheds. Selenium can be as toxic to cattle and wildlife as it has been to cutthroat trout downstream of several coal mines operating on the other side of the Alberta–British Columbia border. Selenium levels downstream of the Teck Coal Limited mines in British Columbia are so alarmingly high that even the EPA under the Trump Administration expressed concern about what they were doing to fish on the US side of the border. The company was eventually charged before pleading guilty for polluting the waters of the Elk River valley. The $60-million fine levied in the spring of 2021 was one of the largest of its kind in Canadian history.

"The coal mines are going to wreck this country," said Blades. "Everything we have here—cattle, bears, wolves, elk, and trout—depend on clean clear water flowing out of the mountains into those meadows. If we lose that, we lose everything. And there will be nothing left but scars on the mountainsides."

Ranchers here have been duking it out with government officials ever since the forest reserve was created in 1910. The aim was to protect valuable timber and the headwaters of the Oldman, Livingstone, and Crowsnest Rivers.

Keeping cattle out of the forest reserve proved to be difficult because there were no fences and few rangers to escort the animals back to where they came from. It also made little economic sense, because ranching at the time was generating more revenue than the felling of trees.

Abraham Knechtel, the federal inspector in charge of the forest reserve, finally figured it out in 1910 when catastrophic wildfires ripped through the region, burning up dry tallgrass that would have once been chewed up by bison.

"It does not seem to me that the department should exclude from the reserves all land suitable for grazing," said Knechtel, who was one of the first Canadians to obtain a forestry degree at Cornell University in the United States. "The forest reserves are for the use of people; then why should good grass be allowed to go to waste if it can be utilized?"[6]

Knechtel understood that cattle could do what bison did before they were driven to near extinction.

"The grazing may be reliable as a protection for the woods," he wrote. "In some places, the ground is covered with a dense growth of long grass and peavine. This, when dry, offers much fuel for fires; and when the fire once gets into it, it is almost impossible to check the flames. Cattle have much the same habit as buffalo. In going to the water, they follow one another and make paths which they follow day after day. These paths are fire lines where the fire may be checked, small to be sure, but there are many of them, and they give lines for which to back-fire."[7]

No one really knows the full extent of the damage that might be done by the coal companies, because there is no comprehensive inventory of groundwater, fens, and wetlands on the Canadian side of the Rockies. Nor is there a benchmark for mosses, liverworts, and hornworts, or for insects such as butterflies and moths and beetles that facilitate fungi in their role in decomposition.

It is dry on the eastern slopes of the Rockies. But there are fens, like Sibbald, 45 miles west of Calgary in the Kananaskis region of the

Alberta Rockies. It is also small—just five square miles in size. Like most mountain fens, it acts like a sponge during the spring runoff and then as a water storage tower when drought dries things out.

This fen played a critical role in June 2013 when a heavy-rain-on-mountain-snow event resulted in a catastrophic flood that tore through the cities of Canmore and Calgary. Had peat and beaver dams at Sibbald fen and other mountain valley wetlands not been there to sponge up and store some of the floodwater, the damage would have been worse.

The proof of this is in the research that scientists and students at the University of Saskatchewan have been conducting at Sibbald fen, and in the long-term data that John Pomeroy and his colleagues have been collecting at Marmot Creek, southwest of Sibbald. Pomeroy is director of the Global Water Futures program and the University of Saskatchewan Centre for Hydrology. Global Water Futures has a budget of about $330 million, funding a network of 190 professors and some 900 graduate students, researchers, and scientists working with almost 500 partner agencies and institutions across Canada.

According to Pomeroy, had the forest soils, mosses, and trees not been there during the flood of 2013, the flood peaks would have doubled. This, he told me, is an unrecognized vulnerability for Canadian Rockies water management should severe forest fires sweep across the headwaters. Such events are much more likely with the warming climate, extreme droughts, and declining snowpacks experienced in the Mountain West of North America in the last decade.

It's become increasingly apparent to Pomeroy that the hydrology of this mountain region is changing due to climate change. Rain rather than melting snow is now the main source of moisture at lower elevations. It is coming down more erratically, with longer and more severe wet and dry spells. While winter minimum temperatures in Marmot Creek have risen by about five degrees Celsius (9°F) since 1962, and snowpacks at lower elevations are now only half of what they were, Marmot Creek streamflow continues within its natural range.

Pomeroy sees a much different scenario unfolding over the next sixty years, when there will likely be a shift to much earlier and flashier streamflow due to lower winter snowpacks, more rainfall in the spring, and longer low-flow periods in relatively drier summers. He suggests that the water-management techniques developed over the twentieth century will no longer be useful in this vastly changed water future. Every fen, tree, and moss-banked stream will count—even more than they have in the past.

Ranchers, he said, are justified in being so concerned about open-pit mountaintop coal mines. "The venality and foolishness of expanding coal mining in the mountains during a time of rapidly increasing greenhouse gases in the atmosphere is breathtaking. Climate change means that the world, including Alberta, is going to face more severe droughts and floods and more wildfire. Heavy rains and rain-on-snowmelt events can mobilize open-pit mine waste as it did at Mount Polley in British Columbia, where retaining dams were overwhelmed. They were not designed for or maintained to withstand the extreme hydrology of the twenty-first century.

"We're going to have to make some decisions about the future," he said. "Will we use our precious water resources for mining coal, or for food and ecosystems? Balancing water use amongst ecosystem and drinking-water needs and irrigation and hydroelectricity will be challenging enough under climate change. The prairies do not have the luxury of a Great Lakes system that can provide a bountiful and seemingly endless supply of water. What water we have, we get from the mountains."

The day after Pomeroy and I first met and hiked up the mountain to visit his research site at Marmot Creek, I did a side trip into Simpson Pass in Banff to retrace a trip that Abraham Knechtel and American climber Walter Wilcox did separately more than a century ago. Knechtel went there to collect heather that was to be pressed and placed into 2,000 pamphlets promoting the establishment of the Dominion Parks system—"A Sprig of Mountain Heather: Being a Story of the Heather and Some Facts About the Mountain Playgrounds of the Dominion."

Wilcox was on his way up to do some botanizing before circumnavigating Mount Assiniboine. He was enamored by the beauty of a "meadow, swamp, and grove" he found along the way. He was especially enchanted by a calypso orchid he spotted growing out of peat. There is, he wrote, "a certain regal nobility and elegance pertaining to the whole family of orchids, which elevates them above all plants, and places them nearest to animate creation."[8]

You don't necessarily have to believe in God, as Wilcox did, to appreciate what he meant by that. When I arrived at that so-called swamp, snow had begun to fall. I had nearly given up looking for the orchid when a ray of light broke through the stormy sky and lit up a calypso in the mossy turf. It reminded me of the folly of our seeing limits in marginal landscapes such as this one. We can expect the unexpected only if we look, as Wilcox did in in 1895 and as John Muir did before him in 1864, while slogging through a bog in Canada, looking for that same orchid.

Muir was reduced to tears when found his calypso. "When the sun was getting low and everything seemed most bewildering and discouraging, I found beautiful Calypso on the mossy bank of a stream, growing not in the ground but on a bed of yellow mosses in which its small white bulb had found a soft nest, and from which its one leaf and one flower sprung.

"The flower was white and made the impression of the utmost simple purity like a snowflower. No other bloom was near it, for the bog a short distance below the surface was still frozen, and the water was ice cold. It seemed the most spiritual of all the flower people I had ever met. I sat down beside it and fairly cried for joy. It seems wonderful that so frail and lovely a plant has such power over human hearts."[9]

It may not have been the orchids that won the hearts of so many people in Alberta who rose up and forced the government to pause for a short time and reconsider its mountain coal policy in the spring of 2021. But somewhere along the line, many of them must have been touched in some way by these tiny mountain fens that play such an oversized role in nurturing so many rare plants and animals. The pity of it is that many of these people continue to be drawn more by the majesty of the mountains and the rivers than by those tiny, mushy fens filled with mosses, liverworts, and hornworts and by blueberries, huckleberries, and other plants that thrive in acidic soil.

Chapter 9

Ring of Fire: The Hudson Bay Lowlands

An ingenuity too astonishing
to be quite fortuitous is
this bog full of sundews, sphagnum-
lines and shaped like a teacup.
A step
down and you're into it; a
wilderness swallows you up:
ankle-, then knee-, then midriff-
to-shoulder-deep in wetfooted
understory, an overhead
spruce-tamarack horizon hinting
you'll never get out of here.

— Amy Clampitt, "The Sun Underfoot Among the Sundews"[1]

In the early 1950s, three plans were put into play to blow up vast peatlands in Alaska and Canada with nuclear bombs. One came from Los Angeles–based Richfield Oil, which got initial approvals from the US Congress and Canadian government to detonate an eighteen-million-pound bomb in the northern Alberta oil sands region to produce the intense heat needed to separate oil from bitumen. Company scientists expected that it would produce a crater large enough to fill several million cubic feet with melted oil. A Canadian government official was confident that blasts such as this would double the world's petroleum reserves.

While studies were supposedly being done to reassure the Canadian public about radiation, groundwater contamination, and earthquakes, American defense officials considered the possibility of testing thermonuclear devices in Alaska for military purposes. The idea was ruled out because of the high costs and the possibility of killing "wandering groups" such as trappers, prospectors, and indigenous people.[2]

Canadian defense officials weren't as concerned when they offered the British the opportunity to detonate twelve Hiroshima-sized bombs in the Hudson Bay Lowlands, a 100,000-square-mile expanse of peatland that extends nearly 900 miles from the province of Quebec to Manitoba. Six percent of the northern hemisphere's peat is found here at depths of up to fifteen feet. These peatlands are rivaled in size only by the vast network of peatlands that cover most of western Siberia.

"The Technical Feasibility of Establishing an Atomic-Weapons Proving Ground in the Churchill Area" was the brainchild of C. P. McNamara of Canada's Defense Research Board and William George (later Lord) Penney, a scientist who participated in the Manhattan Project.[3] They decided on the Hudson Bay Lowlands over several other remote places in Canada because they considered it "valueless"—"a wasteland suitable only for hunting and trapping." Only the "occasional hunter or trapper," would be affected, they reasoned in a proposal that would have had the highest level of attention within the two governments.[4]

Canada entered the nuclear age in 1942, when the Allied cause was faltering following the surrender of France and devastating losses on the shores of Dunkirk and the fields of Flanders. British scientists had made significant progress in determining the feasibility of an atomic bomb by then. But they and the Americans needed a reliable source of uranium oxide to build one. C. D. Howe, a senior minister in the Canadian government, offered a helping hand when he ordered the owner of a shuttered mine on the east shores of Great Bear Lake in the Arctic to resume production in order to produce sixty tonnes of uranium oxide. "Get together the most trustworthy people you can find," he advised. "The Canadian government will give you whatever money is required. . . . And for God's sake, don't even tell your wife what you're doing."[5]

The testing of the bombs in the Hudson Bay Lowlands was to be centered in the Broad River area, 12–45 miles south of the small coastal community of Churchill, Manitoba, and dead center of what would become Wapusk National Park. It was and still is mostly virgin peatland, home to the seven First Nations of the Mushkegowuk Council, who have genuine concerns about the so-called Ring of Fire that is potentially one of the biggest mining opportunities in Canada in over a century. In 2021, there were approximately 13,296 active mining-claim units held by eighteen companies and individuals, covering approximately 800 square miles.

I have visited the Hudson Lowlands seven times over the years, each time with a different purpose and a greater appreciation for just how important this peatland ecosystem is, and for what would have been lost if the bombing had taken place and what will be degraded if the mine developments proceed.

The first three trips were spent in a tiny, uninsulated wooden hut perched on top of a bird tower along the west coast of Hudson Bay at Cape Churchill. Hundreds of polar bears end up at the cape each summer because that is where the last remaining pans of melting sea ice force them to go ashore.

It's a cold part of the world, colder than many places much farther north. When I was there in the late autumn months, there wasn't a day without blowing snow. Temperatures routinely dipped to 20–30 degrees below zero Celsius (-4 to -22°F)—without the wind chill. There was no outhouse, just an outdoor pot on a narrow balcony that shook thunderously in the wind.

There were never more than three of us stationed at the tower because that was all the room there was in the hut. The aim was to detect and deter some of the estimated 1,000 bears in the region before it became necessary to kill them, as miners, offshore oil workers, soldiers, and even wildlife officers often did in the Arctic when they found themselves confronted by one of the more aggressive animals.

Each morning we would climb down the ladder, careful to make sure there wasn't a polar bear in the vicinity. Once we were sure that the coast was clear, we would fire up a large barbecue to cook rancid seal and walrus fat provided by Inuit hunters living up the coast. The putrid smell of smoke drifting across the tundra quickly livened up what seemed like an empty landscape. As if by magic, several bears would pop their heads up from behind snowdrifts, sniffing the air as if it were coming from sweet, hickory-smoked bacon frying in a pan.

Once a bear got within range, the scientists would test a number of strategies to frighten it off sufficiently so that it would not return. Fireworks and amplified heavy metal music only made them curious. The electric fences didn't seem to have the necessary power to deter them from returning. Rubber bullets were unreliable in that they often misfired. Plastic bullets aimed at the animal's posterior finally did the trick. Almost none of those bears came back.

We saw few females because those with cubs tended to avoid males. Those that were going to give birth were already twenty to fifty miles inland in the Broad River and Owl River areas, nestled up in dens that that had been excavated into the side of peat banks along a bog or lake.

I got a chance to see these denning sites many years later when I participated in a survey of the Broad River and Owl River area while the females were emerging and heading back to the coast with their cubs to resume the hunt for seals—food they hadn't consumed for several months. The fast begins in June or July, when the melting sea ice forces them onto land. It ends in March, when the cubs are old enough to make the twenty- to fifty-mile trek back to the coast.

It was odd seeing polar bears plowing through a snowy forest with a cub or two in tow; the stereotypical image of a polar bear is one that walks effortlessly on ice and dens in snowbanks. The bears here and those in neighboring southern Hudson Bay are unique in that they have a long-term fidelity to these peatlands. They've been digging dens in peat banks for at least several hundred years and possibly longer.

There are more than 1,200 polar bear dens in the Hudson Bay Lowlands, not all of them used.[6] At any given time, up to 190 females will occupy them.

Partway through that three-day survey, I asked Daryll Hedman, the man who oversaw polar bear management in western Hudson Bay, if I could explore one of the dens. He was happy to oblige after the helicopter pilot did a few flyovers to make sure the one we picked was no longer occupied.

Above: For hundreds of years and likely longer, female polar bears have been denning in the sides of peaty hillsides of the Hudson Bay Lowlands, the second-largest peatland in the world. (Photo: Edward Struzik.) Below: The author emerges from a den.

The den had a just a single chamber. Others have two—a big one for the female and a much smaller one for cubs.

I thought it would be easy to enter the den. But this one turned out to be a lot snugger than I expected. It was more of a crawl and then a roll-over than a walk-in situation. Inside, the glow of my flashlight illuminated a well-worn carpet of moss and lichen that the female had raked in. Roots of birch and larch were growing inward. I did not notice that it was much warmer inside, but it was easy to imagine the temperature rising to more comfortable levels when a 300- to 400-pound bear and one or two cubs were squeezed in.

The animals were not alone out there. When scientists Peter Scott and Ian Stirling conducted a survey of this specific denning area, they found 31 dens.

It was surreal to think that everything within 500 yards of the bombing sites would have been "lethal for a day or two, and that only after several weeks would a rapid decay of radioactivity" allow specially garbed scientists to enter the area for short durations. Ground zero—the point of detonation—for each successive bombing, they noted, would have to be at least three miles away from previous blasts.

It's hard to know what McNamara was thinking when he insisted that no one be told ahead of time. No mention was made of evacuating three of the seven Mushkegowuk First Nations who lived, hunted, and trapped in the region. "Every effort must be made to keep secret the nature of the trial before the event," according to his report. "Once detonation has occurred, there will be little hope of keeping secret the fact that an atomic explosion has taken place."[7]

Polar bears would not have been the only victims. I was mindful of that when we continued with the denning survey. Along the way we saw countless numbers of muskrat lodges. In one case, a lynx was hovering over the top of a "push-up" that the muskrat uses and maintains in order to emerge from the icy depths in winter. The lynx was quivering with anticipation—too hungry to abandon a potential kill even with the giant whirlybird passing by.

Farther along, and flying low, we spotted all kinds of tracks in the snow: a pack of wolves following a polar bear and two cubs; an Arctic or red fox following the wolves. There were signs of moose and hares moving in all directions. It was impressive to see so much activity in this so-called wasteland. And then without notice, a herd of about 700 caribou suddenly came into view, dashing off in the other direction as a cloud of dusty snow billowed in their wake. It was like an African Safari in deep freeze.

The Hudson Bay Lowlands is a much different place in late spring and summer, when its big rivers are flowing—the Seal, Churchill, Nelson, and Hayes in Manitoba, the Severn, Winisk, Albany, and Abitibi in Ontario, and the Eastmain and La Grande in Quebec. On a field trip I did with a biologist from the Canadian Wildlife Service, I learned that the Hudson Bay Lowlands have the highest proportion of palm warblers and yellow rails, a chicken-like marsh bird that is rarely seen, little studied, and very mysterious. The bird, according to ornithologist Alexander Sprunt, is more like a "feathered mouse" than a bird because it seems to prefer to run and hide rather than fly.

During migration, few other places in the world have this many red knots, Hudsonian godwits, ruddy turnstones, black scoters, pectoral sandpipers, semipalmated sandpipers, white-rumped sandpipers, greater yellowlegs, and lesser snow geese. The Hudsonian godwit, which was once regarded as one of North America's rarest birds, stands out among them because it has made a notable, if shaky, comeback. Two-thirds of these birds, some of which fly nonstop from their wintering grounds in southern South America, stop over in their migration along the Hudson Bay shoreline. Many of them breed in the Hudson Bay Lowlands. This is why the Audubon Society is in full support of the Mushkegowuk Council chiefs who are calling for the establishment of an indigenous-led Marine National Conservation Area to protect that breeding habitat, the denning grounds of the polar bears as well as the caribou that dwell in the peatland regions.

If there is a peatland outside Siberia that the Hudson Bay Lowlands can be compared to, it's the labyrinth of fens, bogs, and swamps in Polesia, which stretches out from the riverbanks of the Bug in Poland, the Dnieper in Ukraine, and the Pripyat in Belarus. The comparison to Polesia is apt only because the mires there are also among the last refuges for globally threatened birds such as the aquatic warbler and also for wolves, lynx, bison, and other animals that are threatened throughout Europe.

Polesia's Almany (or Olmany) Mires is one of the largest intact peatlands in Europe outside Scandinavia. But it has been degraded by power lines, border control infrastructure, the Chernobyl nuclear disaster, and a seventy-mile network of forestry roads that the government of Belarus has recently sanctioned.[8]

In contrast, the Hudson Bay Lowlands are more than ten times the size and largely unexploited except for hydroelectric dams and a handful of mine sites. The meltdown of the Laurentide ice sheet is so recent that land pushed downward by the enormous weight of that ice is still rebounding, reshaping the landscape and reconfiguring the drainage in ways that favor the growth of marsh communities of aquatic sedges before sphagnum and other mosses take over and turn them into a bog or fen.

Paludification is the hydrologist's word to account for this kind of ecological succession. There is no other direction for rain, melting snow, and river floods to go but sideways on impermeable, finely textured silt and clay that is as flat and frozen as it is in this region. Peat not only continues to grow here, it insulates the permafrost below and stores vast amounts of carbon that would, if unleashed, warm the world much faster than it is already warming.

I never thought much about the implications of this until 1990, when I crossed paths with scientist Nigel Roulet. Roulet and more than two dozen other scientists were collaborating with NASA to sample the chemistry of the atmosphere above the Hudson Bay Lowlands. It didn't occur to me back then that the carbon stored in peatlands could be a major driver of climate change if it was disturbed or thawed out of permafrost areas. Chinese rice paddies, belching cows, and dirty diesel trucks were getting most of the media's attention in those early stages of the climate-change discussions.

Scientists now know far more than they did in 1990 about how much carbon is freed when permafrost thaws, when trees and shrubs growing on top of it burn, or when it is disturbed in other ways. Where that thawing is in high gear, as it is in Tanana Flats in central Alaska, and in the Scotty Creek area of the Northwest Territories, it is dramatically reshaping the landscape, reversing the succession in some cases from marsh to fen and forest and then back to fen, bog, and marsh.

At Tanana Flats, permafrost degradation increased the size of three large fens by 26 percent from 1949 to 2018. Torre Jorgenson, the lead author of a report on the degradation, says that the trend to warmer, snowier winters (snow traps warmth) has pushed the region past the tipping point where permafrost will no longer form and where the complete thawing of permafrost is inevitable.

Canadian scientist Bill Quinton and colleagues have seen the same thing playing out in the sub-Arctic region of Scotty Creek in the Northwest Territories. Scotty Creek drains about sixty square miles of fens and bogs. Where there is permafrost at Scotty Creek, it is warm and vulnerable to thaw. This is significant because snow acts like a blanket, trapping some of the heat that has thawed the ground in summer. Thick, long-lasting snow cover followed by a quick spring meltdown can hasten the thawing of the frozen ground in places like this. (In areas where permafrost is meters thick and rock-solid, it insulates the ice from the warming rays of the sun.)

Snow measurements at Scotty Creek began in 1994, five years before Quinton set up a semipermanent field station. It is rare to have long-term data like this in the Canadian North, one that underscores unequivocally what happens when winter snow is deep, spring runoff is significant, and the thawing of relatively warm permafrost switches into high gear. It's why so many scientists have come knocking on Quinton's door, asking to come in and participate. It's also why the Dehcho First Nations, which will one day become the legal owners of the area, are keenly interested and actively participating in what he is doing.

The peat at Scotty Creek is typically nine to twelve feet deep, and thirty feet in some places. In the 1950s, permafrost covered nearly three-quarters of the region. It's down to a third of that. The edges of those local patches of permafrost are receding by about three feet each year.

Trees are literally drowning in depressions created as the permafrost thaws and the ground surface collapses, and as melting snow and rain fill them up with water. The transition from land to water has been so dramatic that Quinton has had had to move base camp twice. "It's crazy. There is water everywhere," Quinton told me. "What we're seeing perhaps more clearly than any other place in the world is ecosystem change occurring in fast motion. The implications for water quality, vegetation changes, biodiversity, and the people living in this part of the world are profound. Many of them are indigenous hunters, trappers, and fishermen. They tell me they have never seen anything like this."

The best way to visualize what Quinton is talking about is to describe what occurred 200 miles to the northeast in the Mackenzie Bison Sanctuary, where warming temperatures and wildfire have thawed and collapsed peatlands in and around the sanctuary so intensely that incoming water drove most of the 700 wood bison out of the protected area. Some 385 fires burned a record 3.4 million hectares (8.4 million acres) of the Northwest Territories in 2014. The exodus was so complete that Terry Armstrong, a biologist working for the government of the Northwest Territories, had a difficult time finding animals when he flew in to do a count in 2015.

No one can say with certainty whether it was paludification that drove the animals out. But York University scientist Jennifer Korosi says that it's hard not to make the connection, considering that the amount of water in the sanctuary doubled from 1986 to 2014. Falaise Lake, the biggest body of water there, had grown in area by 824 percent.

"The short answer is that we do not know exactly what's behind this," Korosi told me. "But the tree ring and sediment coring that we did suggest that the hydrological changes that are taking place are unprecedented in the last 300 years. The area has flooded before, but nothing close to the scale that is now occurring. This is backed up by the Dene elders who hunt and trap in the area."

There is some debate as to whether this thawing of permafrost in peatland ecosystems is, as Canadian scientist Antoni Lewkowicz describes it, "a long freight train moving slowly as it unloads its greenhouse gases," or as others warn, a carbon bomb about to blow as the train careens off the tracks. In the Hudson Bay Lowlands, it's more of a slow-moving train, which is fortunate because the Lowlands store 12 percent of all the organic carbon in Canadian soil.

Canadian scientist Elyn Humphreys and her colleagues monitored two peatlands sites in the James Bay lowlands close to the Victor Diamond Mine site, which is located in the so-called Ring of Fire.

One of the peatland sites Humphreys studied is dominated by evergreen vegetation. The other is dominated by deciduous plants. Each one is responding differently to warming temperatures. In each case, the peatlands are absorbing less carbon dioxide as temperatures rise. But the amount of methane being released is still less than 10 percent of the carbon dioxide that is being taken up.[9]

It's a similar situation in the Mer Bleue Bog, which is located seventy miles to the southeast, outside the city of Ottawa. Nigel Roulet had been monitoring this for about as long as Bill Quinton has been conducting research at Scotty Creek. When he and his colleagues started out in 1998, they assumed that it wouldn't take much warming for Mer Bleue to dry out and release carbon dioxide and methane. But as scientists like Humphreys, Tom Moore, Lorna Harris, Paul Wilson, and others got involved, it was clear that there is a resiliency in this self-regulating ecosystem that is keeping most of the carbon in the ground.

It's a different story, however, in peatlands like those in the Bonanza Creek region of Alaska. Many of the more than sixty scientists who have been conducting research there have observed how quickly the boreal forest is browning, how wildfires are burning bigger and more intensely, and how conifers that burn are being replaced by deciduous trees. In some places, methane emissions have tripled. In others, they have remained fairly stable. There is a lot to account for the discrepancies and the uncertainties. Temperature, precipitation, wildfire, snow cover, vegetation, and even soil chemistry and beaver dams can profoundly affect the hydrology and the amount of carbon dioxide and methane that is released.

The sometimes bewildering weight of evidence on each side of the carbon bomb–slow-moving freight train debate reminds me of the seesaw nature of Arctic sea-ice predictions in the early 2000s.

No one back then envisioned that the Arctic would be as ice-free as it was in 2021. This increasingly ice-free Arctic is not only changing the temperature and chemistry of the Arctic Ocean, it is also opening the door for southern species such as killer whales and Pacific salmon to move in and kill or overtake beluga, narwhal, and Arctic cod.

Even if this thawing of the Arctic and sub-Arctic is just a slow-moving freight train, as it no doubt is in most permafrost regions, it is still a major concern because there is so much carbon stored. Northern peatlands cover an estimated 1.4 million square miles, an area larger than all of the western United States and Texas combined. They have accumulated as much carbon as there is stored in all of the world's forests combined, half of it stored in permafrost. A one-degree increase in temperature has the potential to free up the equivalent of four to six years of fossil-fuel emissions.[10] This is all spelled out in 2020 report produced by Swedish scientist Gustaf Hugelius and thirteen scientists from around the world, who concluded that global warming may very well transform northern peatlands from landscapes that cool the climate to ones that will eventually warm it.[11]

If the situation in the West Siberia Lowlands represents what can be expected in the Hudson Bay Lowlands in the future, it's an unnerving scenario. The West Siberian Lowlands extend 1,200 miles from the Ural Mountains in the west to the Yenisey River in the east. The length from north to south is 1,500 miles. It's seven times the size of Germany and about a third bigger than the Hudson Bay Lowlands.

At the turn of this century, most scientists viewed the Siberian Lowlands as a relatively reliable carbon sink; it was storing more carbon that it was releasing into the atmosphere. But scientists didn't really know this for certain, because basic information about peatland age, thickness, and hydrological factors was sparce. It's clear now that higher temperatures, permafrost thawing, wildfires, and drilling for oil and gas have destabilized the hydrology, the nutrient cycling, and the vegetation cover. Parts of lowlands of Siberia are not so much a carbon bomb as they are a volcano rumbling before an eruption.

In just one month in June 2019, fires in Siberia sent five million tons of carbon dioxide into the atmosphere. That was equivalent to the total amount of carbon the country of Sweden emits in a year.[12] It was even worse in 2020, when the amount of carbon released by fires increased by 35 percent. About half of the carbon came from burning peat.

What's happening in Russia in not surprising, because the history of fire there in this young century is akin to what has been happening in California more recently. Five of the top twenty fires in the state's history occurred in 2020. Since the record-breaking Siberian fires of 2003 burned more than 77,000 square miles (200,000 sq. km) of forest, grasslands, bogs, fens, and marshes, the number of big peat fires has been escalating.

No one thought that anything like this could happen in the tundra regions of the North American Arctic, because it was considered to be too cold and wet in summer for fires to ignite and spread and sustain themselves. It's why scientists based at the Toolik Field Station on the North Slope of Alaska paid little attention when they smelled smoke in the summer of 2007. They figured the smoke was coming from fires burning in the boreal forest to the south until one of them noticed that the smoke was coming from the north.

The Anaktuvuk tundra fire ended up burning 401 square miles (1,039 sq. km). The fire released 2.3 million tons of carbon into the atmosphere. A tundra fire of that size had not happened in the past 11,000 years.[13] Not only are these peat fires contributing to climate change, they are also releasing mercury into the atmosphere at up to fifteen times the rate of upland forests. Even the soggy peatland forests in the Hudson Bay Lowlands have not been spared. Polar bear den sites have been destroyed by fire, and bears are now avoiding the burned areas.[14]

Another catalyst that accounts for rapid thawing is the bacteria and archaea (single-celled organisms that thrive in low-oxygen environments) that produce carbon dioxide and methane. The more unstable the active layer of permafrost becomes after a fire, flood, or warm winter, the more carbon there is available for many of these carbon- and methane-producing microbes. Long-term studies in peatlands such as Sweden's Stordalen Mire have clearly shown that it is no longer famine for microbial communities in discontinuous permafrost, but rather a feast that is further accelerating thawing, altering plant habitat as well as the nature and structure of methane-producing microbial communities that awaken the organic carbon. What no one is certain of just yet is whether this microbial cycling is destined always to release carbon in rapidly thawing permafrost—or whether there might be a recovery in peat accumulation once the ecosystem stabilizes and starts growing moss once again.

The British never did detonate those atomic bombs in the Hudson Bay Lowlands. It wasn't so much the cost ($62 million in 2021 dollars), nor was it the need to spend another $45 million on a railway spur line from Churchill, Manitoba, to the Broad River area. Instead, the Brits decided to detonate their bombs in Australia's Montebello Islands because they feared that northern Canada would be too cold for the 350 scientists, technicians, and military personnel that would have been stationed there. It was a good decision for Canada, because even today, visitors to Montebello's ground zero sites are advised not to spend more than an hour there and to avoid touching anything because it could still be radioactive.

Bombs did blow up in northern Canada, though. Throughout the Cold War, the Hudson Bay Lowland outside of Churchill, Manitoba, was used as an artillery-training grounds as well as a test site for missiles and rockets. As many as 171 rockets were launched annually by Canada's Defence Research Board and the National Research Council.

The last time I was in the Hudson Bay Lowlands, I was once again participating in a polar bear survey. This time we were counting the number of animals that had come off the melting sea ice. It was enthralling, as the four of us in the helicopter counted aloud the animals we were seeing: a female and two cubs in two feet of water feeding on the carcass of a beluga whale; a celebration of eight big males sitting on a sandy spit, ignoring each other when they would otherwise be fighting over females or food if either were available. A black wolf, bloodied by a small pack that was following it, took no interest as we hovered overhead. It seemed to be more afraid of what the wolves would do to it than what we might do. Our flight ended with the fattest bear I've ever seen heading south into the boreal forest to find a den in those peatlands.

The excitement of that last trip didn't end there. Back at my home office, I got an e-mail from Pierce Roberts, Daryll Hedman's boss, with a photo of a grizzly bear standing by a polar bear it had just killed and partially buried. It was extremely rare at the time to see a grizzly bear in this part of the world, and unheard-of to see one that has just killed a polar bear.

After I filed the photo into my computer's library, I began sorting through the pictures I had taken on that trip. While scanning through the aerial shots of the polar den I had entered along the hillside of that fen, my eye was drawn to something in the forested background that didn't look right. As I magnified the image, I could see it was a polar bear sitting (hiding?) behind a tree and wondering, I could only imagine, what I was doing in its den.

Chapter 10

Pingos, Polygons, and Frozen Peat

We hoped a breeze might scatter the mosquitoes, but when we dragged our tent and bedding up the muddy bank on the muskeg, they greeted us with black swarms. We crawled under our mosquito bars fully dressed, kicked off shoes and gaiters, took off hats and veils and went to sleep on a bed of Ledum, moss and cranberry.[1]

—Australian social worker and travel writer Clara Coltman Vyvyan on a torturous 600-mile canoe trip she made in 1926 with Gwendolen Dorrien-Smith down the Mackenzie River, up the Rat River, and down the Porcupine River with two Dene men guiding them two-thirds of the way to Alaska

As the sun in the western Arctic descended lower on the horizon, the motley, late-summer colors of the tundra began to light up. A painter's palette of early September: rusty oranges, butterscotch yellows, and reds ranging from crimson to carnelian, all swinging and shimmering in the steady breeze. It was autumn in Vermont, but without the trees. At this time of year, the warm, vivid hues take over from the cool greens when the temperature drops, breaking down chlorophyll in the leaves of blueberries and cranberries, willow, alder, bog birch, and many other shrubs that thrive in frozen peatlands. The surprising persistence of cotton grass (*pualunnguat* in Inuktitut), continuing to bloom this late in the season, added a few splashes of white to the color scheme.

It was cool, as one would expect north of the Arctic Circle. The prospect of rain, or possibly wet snow, presented itself in the leaden clouds that hovered over the Beaufort Sea. Nothing cuts to the bone like the wet, wind-driven polar weather that blasts through the Mackenzie River delta in late summer and fall.

A storm was not a matter of concern to me because I was in a warm, mud-caked pickup truck driving bumpily from the small town of Inuvik (population 3,200) in the Northwest Territories of Canada to the even smaller town of Tuktoyaktuk (population 700) on the only public all-weather road on the continent that goes to the Arctic Ocean. (*Tuktoyaktuk* is the Inuit, or Inuvialuktun, word for "looks-like-a-caribou.")

Philip Marsh, a cold-climate water scientist at Canada's Laurier University, suggested doing this drive the day before we were to fly into his field camp in the Trail Valley Creek region, where relatively thin, fragmented layers of peat sit on top of permafrost that is up to 1,500 feet deep.

It's part of a much more expansive area of permafrost—ground that stays at or below 0 degrees Celsius (32°F) for at least two years in a row—covering nearly nine million square miles in the northern hemisphere.[2] About 1.4 million square miles of peatlands within those areas of permafrost and those that freeze seasonally in the northern hemisphere are thawing, sometimes abruptly.[3]

Marsh wanted to get a firsthand look at the highway concept he and others worked on nearly forty years ago when environmental and hydrological assessments for a road to Tuktoyaktuk were first put into play. Literally a pipe dream at the time, and for many decades after, the two-lane, 87-mile gravel road was meant to be the final section in the Dempster, Klondike, and Alaska highway systems. It finally became a reality in the late fall of 2017, when it was officially opened to traffic with much fanfare from newspapers around the world.

The road was billed as an engineering marvel. Recognizing that peat acts as an insulator, builders chose not to remove it when they laid sand and gravel on top of the permafrost. Rather than have the road block the flow of rain and snowmelt that would pool and warm the permafrost by the roadside, culverts were installed to allow water to flow through. That in itself is not unusual; but installing 300 of them for such a short road like this one was very unusual. Construction was limited to the winter months so that the underlying permafrost would not be disturbed.

Costly as it was at a little over $300 million, there is no guarantee that the road will be able to withstand the abrupt thawing that is taking place in the western Arctic. The distribution of ground ice is difficult to measure, and it is even more difficult to predict how it will behave as precipitation patterns change in a region that is warming three to four times faster than most anywhere else. Permafrost has a distinct personality that is not easily understood, according to Marsh. There is something new to be learned about it with each year that passes.

The complex, taciturn personality of permafrost has become a nightmare for those road engineers who have spent as much as $3 million shoring up just 500 yards of the 1,390-mile-long Alaska Highway. If the road to Tuk collapses, as it has begun to do less than three years after it was opened to traffic, it will underscore yet again how little we know about tundra regions, where the thinnest layers of peat in some places are insulating massive amounts of frozen water that are keeping everything glued together.

A few miles north of Inuvik is where the tree line turns to tundra and where we got our last look at the Mackenzie River. To the unaccustomed eye, this landscape is as close as it gets to MAMBA country, the acronym Welsh farmers use to describe the undulating blanket bogs of the Elan Valley where there are "Miles And Miles of Bugger-All." To the credit of British author Robert Macfarlane, there is a whole glossary of Cornish, Welsh, Irish, Gaelic, and other words that infuse it with more value.[4]

It's a pity that the nineteenth-century British explorers who came to this part of the world searching for an overland route through the Northwest Passage were indifferent to, or unaware of, the wonderfully descriptive language that Macfarlane has helped to rescue in his book *Landmarks*. Consider *foggit*—the Scottish word for ground covered in moss and lichen, or the *bull-pated* landscape we saw all around us along the roadside. It is a term that applies to wind-swept tufts of grass.

Unable to fit these vast vistas into the familiar schemata of the British landscape, the explorers often dismissed the tundra as "sterile" or "barren." "Amongst these hills you may observe some curious basins," wrote surgeon-naturalist John Richardson in a letter to artist and fellow officer George Back in 1821. (Richardson and Back were members of the first of three expeditions led by John Franklin, the hapless explorer who disappeared in 1845 with 128 men while searching for a sea route through the Northwest Passage.) "But nowhere did I see anything worthy of your pen. So much for this country. It is a barren subject and deserves to be briefly dismissed."[5]

There was much more to Richardson than a superficial observation such as this one. Charles Darwin was an admirer and queried him on a number of Arctic matters, including how far north the forest grew and what kind of trees and plants rose up from frozen soil.[6] John James Audubon, Louis Agassiz, Richard Owen, and other luminaries were friends. "Plants of northern Canada named for Richardson would make a garden of respectable size, and animals named by and for him a considerable zoo," noted a Canadian admirer.

Richardson was no slouch. On the first expedition, he tried to swim across the icy Coppermine River to help save his starving men. On Christmas Day, during the second expedition in 1826, he embarked on what may have been the longest, most grueling birdwatching expedition ever—a 900-mile hike from the icy shores of Great Slave Lake in the sub-Arctic to what is now Fort Carlton in Saskatchewan so that he would be there in time for the spring migration. He was sixty years old when he returned to the Mackenzie Delta for a third time in 1848–49 to search for Franklin.

In Richardson's time, the value of peatlands was measured not so much by their flora and fauna but by the potential of turning them into farmlands and forest and reducing the spread of disease. There was little in the tundra that was picturesque and pleasing to the eye because, unlike the Scots who grew up and hiked and hunted in the blanket bogs of the Flow Country, most Brits were not trained to understand the many virtues of soggy peatlands beyond their reputation as the haunts of will-o'-the-wisps and corpse candles and they were vectors for marsh fevers, malaria, and cholera.

Richardson and his companions paid a price for not fully appreciating the tundra for what it offered—blueberries, cranberries, cloudberries, Labrador tea, the roots of some louseworts (which are lemon yellow and sweet like carrots), the succulent leaves of mountain sorrel, wild potatoes, rhubarb, and celery, and arctic fireweed and purple saxifrage, which makes a delicious tea.

Only nine of the twenty men on the expedition survived. Those who died either starved to death or were killed by a voyageur who was suspected of killing and cannibalizing some of them. Richardson shot him, claiming self-defense when it was, in all likelihood, a summary execution. Those who lived were forced to cook up old deer skins, leather boots, and lichen (*tripe de roche*) that they scraped off rocks.

As the British explorers starved and stumbled, the Inuit were carving out a comfortable existence at places such as Kitigaaryuit, the name of a verdant peatland located on a spit jutting out from the East Channel of the Mackenzie River, not far from where we drove that day. For more than 500 years, the Mackenzie Inuit (Kitigaaryungmiut) gathered at Kitigaaryuit each summer to hunt beluga whales and caribou, pick berries, dig up root vegetables, and gather peat moss for lighting seal oil, covering roofs, and plugging walls. *Ugrruk*, the Inuit word for peat moss, also identifies a plant that can be used for diaper linings.[7]

Those who stayed through the winter lived in elaborate multi-family houses framed with driftwood and insulated with sod (peat). David Morrison, an Arctic archeologist with the Canadian Museum of Civilization, spent a good part of his career exploring long-abandoned sites such as this one. No other Inuit group in Arctic Canada, except those who lived in similar fashion on Herschel Island to the northwest, lived as prosperously, he once told me.

The drive to Tuktoyaktuk ("Tuk") reminded me that there is exquisite beauty in the tundra if you take the time to look. It's not just the shafts of sunlight piercing through dark, leaden clouds that make the autumn hues pop. Nor is it the cold, green, diamond-like sparkle of the northern lights shimmering across the sky at night. Over time, the thawing and freezing of peat and minerals in the permafrost can result in artful, polygonal patterns that are widespread in terrains such as this.

Ice-wedge polygons begin to take shape in winter, when cracks in permafrost form and when the soils, especially those infused with expansive, water-logged peat, shrink under extremely cold conditions.

The Inuit of the circumpolar world, such as these people in Alaska, used peat to construct homes. (Photo: F. H. Nowell / Library and Archives Canada / Edward Martin Kindle Collection / e011303081-052_s1.)

When snow melts in spring, some of the water seeps into these cracks. Ice wedges form as the water is chilled and frozen by the thick layer of permafrost below. When it heaves up in the center, it resembles the back of a turtle. There are many striking examples of this, but none as beguiling as those that are found at Ballast Brook at the north end of Banks Island, where, just a few miles away, there are stumps of 4,000-year-old spruce trees rising up from beds of peat that preserved them.

Up close, the picture can be just as enchanting when pseudomorphs—minerals formed by chemical or structural change of another substance—form as both meltwater and sediment accumulate in the cracks of frozen soil. Depending on the nature of those sediments, the pseudomorphs can appear to be crystalline, even though they are not.

The Tuk Peninsula is best known for 1,350 pingos. *Pingos* are ice-cored hills that form at the bottom of a slope where groundwater is forced up and grows as ice mixed in with peat and sediments. Some pingos are as big as a football stadium. Others partially collapse when the peat is no longer thick or compressed enough to insulate the ice inside. Shallow lakes or ponds sometimes form in the cavities. The massive Ibyuk (Inuvialuit word for sod) Pingo is about 150 feet high, 1,000 feet wide, 1,000 years old, and still growing at about three-quarters of an inch each year. The record for annual growth of a pingo is 34 centimeters, or 13.4 inches.

John Richardson had no idea what a pingo was when he sketched one near here during his unsuccessful search for Franklin. Richardson guessed that the icy hill may have been the product of drifting sand. Others attributed the formation of pingos to volcanic activity. Had they asked the Inuvialuit, they would have learned that not only was this formation a pingo, it was an *aklisuktuk*—a pingo that is growing fast. Vilhjalmur Stefansson, the Canadian-born American explorer, lent credibility to this in 1919 when he wrote that Inuit elders he talked to "have noted changes in the appearance of mounds, even an increase in size during a lifetime."[8]

The fact that the word *pingo* didn't come into scientific use until 1938 is telling. Botanist Alf Erling Porsild, a Dane who grew up in Greenland and spoke the Inuit language, appreciated the word for what it was while engaging the Inuit during his survey of the Mackenzie Delta region.

Pingos aside, many of the early Arctic explorers found the tundra to be distant and disorienting, which is forgivable; even now, after decades of traveling through the Arctic, a sense of scale and depth of field often eludes me. Without trees, buildings, fences, high hills, or mountains to use as reference points, the mind's eye is often confounded by the uniformity of so much open space. Through the lens of a spotting scope, the back of a far-off Richardson's ground squirrel perched on a peaty hummock might be mistaken for a barren ground grizzly bear. (The squirrel, like so many other Arctic animals and plants, was named for explorer John Richardson.) Stefansson made just such a mistake while scanning this landscape for game. When the squirrel/bear suddenly turned and charged, he abandoned the scope, ran for his gun, and tripped, certain just then that he was about to be mauled.

In an article written for *Maclean's* magazine in 1929, Stefansson lamented the fact that perceptions of the Arctic had not changed since Richardson passed through. Students, he complained, were still being told that the farther north you go, the colder it gets, and that the Arctic is constantly being buffeted by blizzards—when, in fact, many Arctic regions are not. Even now when I speak to an educated audience and tell people that the vast peatlands of the Old Crow Flats in the Yukon Territory experience July temperatures similar to those in Seattle, some scoff. They have an ever harder time believing that parts of Ellesmere, the most northerly place on the continent, are home to cotton-grass meadows so thick they could be harvested to stuff pillows, as the British once did in the moors of Sussex.

As much as Stefansson was right, though, he was wrong in suggesting that the North American Arctic was "friendly"—a landscape capable of supporting reindeer and muskox herds in numbers that could help feed the world. The Canadian government learned that lesson in 1929, when they bought 3,000 reindeer in the hopes of turning the Inuit of the Mackenzie Delta into game ranchers.

The government purchased the animals from Carl Lomen, a member of a family that owned a bustling reindeer business in Alaska. Lomen was well known for persuading the owners of New York's Macy's department store to stage an annual Christmas parade led by Santa, accompanied by reindeer he shipped in from Alaska. Because of his notoriety, the Canadian sale was big news. The *New York Times* predicted that the task of getting the animals from Alaska to northern Canada will be "one of the most remarkable treks of animals ever attempted."[9] A writer at the *Montreal Gazette* envisioned the arrival of the herd as the opening of a "new era . . . a meat industry of great proportions" in northern Canada.[10]

What was supposed to be a two-year, 1,500-mile journey herding the animals to the Tuk Peninsula turned out to be a five-year ordeal. Only 15 percent of the original animals were delivered. The birth of calves made up for some of the losses. Andrew Bahr, the man hired to herd the reindeer across spongy muskeg, over frozen rivers and lakes, and into snow and ice storms faced a mutiny at one point when his fellow herders thought he was losing his mind. When he finally made it, only two of the original herders were still with him.

Emotionally and physically wasted, Bahr went home to Seattle to a hero's welcome, "the Arctic's Moses," only to find that the house he owned had been lost to taxes during the Depression years. In his absence, the reindeer did not fare well on the tundra. Grizzly bears and wolves feasted on the animals. The Inuit didn't warm to the idea of being herders. Nor did they discriminate between wild caribou and reindeer when they went hunting.

We saw no sign of reindeer on the drive through the grazing reserve. Nor did we see wild caribou, which was also not a surprise. Like most of the world's caribou herds, the Cape Bathurst herd has been struggling to adapt to a climate-change scenario that favors biting flies, more extreme weather events, and the growth of shrubs that shade out the lichen and sedges that help get the animals through the winter. The Cape Bathurst herd numbered just 2,000 in 2019, down from a high of 10,000 two decades earlier.

Investing in reindeer wasn't the only mistake the Canadian government made in the western Arctic. Its answer to the energy crisis of the late 1970s was the National Energy Program. It was so unsuccessful and frightfully wasteful that Canadians still talk about it today. Oil and gas companies, which had little interest in the Arctic outside of Prudhoe Bay in Alaska, were offered as much as 90 cents for every dollar they spent on frontier exploration. They came in droves, spending so much, with so little restraint, that fresh Alberta steaks and live lobsters from the Maritimes were regularly flown into Tuktoyaktuk to feed the crews. Not enough oil or gas was found offshore or on the tundra to make production economically viable.

Scars left behind by those misadventures are not unlike those fens in Colorado that were degraded by mines and diversion projects. The wounds, however, are different. Typically, in the Arctic, oil and natural gas fill up the spaces in underground fractures and pores. When the oil and gas are pumped out, the fluid pressure is reduced. These open cores gradually close up or collapse. The peat and sediments on top subside.[11] Water from melting snow and thawing permafrost pours into the depressions. The water accelerates thawing of the permafrost.

This is what happened at the Kendall Island Migratory Bird Sanctuary, which is located along the outer edges of the Mackenzie

Delta, seventy miles west of Tuktoyaktuk. The sanctuary was established in 1961 to protect a half-million nesting geese, swans, and many other birds. The last Eskimo curlew, not seen anywhere since the 1960s, once nested here.

Sanctuary is not the best word to describe what the refuge is now. In the run-up to build a $16-billion natural gas pipeline down the Mackenzie Valley in the early 2000s, Shell Canada Limited made an application to establish an oil and gas exploration facility on the island. It included a field camp, a 2000-cubic-yard fuel-storage operation, a 2,000-foot-long airstrip, various buildings, and an area for equipment and supplies.[12]

The Canadian government, guardian of the sanctuary, evaluated Shell's proposal and inexplicably concluded that none of the activity was likely to cause adverse environmental impacts. Many nesting sites are now underwater because of the subsidence that followed when twenty wells were drilled and 1,500 kilometers of seismic line (narrow surveying corridors) were cleared.[13] The number of nesting birds appears to have declined.[14] No one knows for sure because the Canadian government isn't sending scientists in to figure that out.

The vulnerability of peat in a permafrost environment explains why the road to Tuktoyaktuk is, dollar per mile, one of the most expensive ever built. Before this road was built, people got in and out of Tuk by boat, by snowmobile, by airplane, and by an ice road that was groomed on the Mackenzie River. The viability of the ice road, however, came into question as a rapidly warming climate shortened the winter season and made parts of the ice road too fragile to support cars and trucks. With no other economical way of getting in and out of Tuk, calls for a gravel road to the Arctic Ocean picked up steam.

It was Stephen Harper, the Canadian prime minister, who came up with two-thirds of the $300 million needed to build the road. Harper was not so much a sympathetic friend of northerners looking for a cheaper way of bringing in food as he was an unapologetic ally of the oil and gas industry. The son of an Imperial Oil executive, he once described the idea of implementing greenhouse-gas emission regulations as "crazy." During his tenure, Harper bet big on oil sands, which have so far destroyed or degraded 239 square miles of peatland in Alberta.

A road to Tuk, his government boasted, would promote tourism and make it cheaper to transport food and supplies into the community.

Even conservative pundits who worshipped Harper thought the idea was crazy. The *National Post*'s Peter Foster, among the most conservative columnists in Canada, described Harper's Arctic highway as a "road to nowhere." He wondered whether the money would have been better spent on something else.

The real reason for the road was kept largely out of public view in a government document that got little media scrutiny. It estimated that the road "reduces oil and gas company investments costs, which leads directly to greater company cash flows." If the road was built, the report stated, that cash flow to industry would amount to as much as $516 million for energy companies.[15]

The oil and gas industry never paid a cent for the highway project. Nor did it get a chance to capitalize on the windfall. A month after the road was officially opened to traffic, the consortium of companies behind the $16-billion natural gas pipeline that was supposed to send gas from Kendall Island to power the oil sands and other industries down south, abandoned the idea. The only people driving the road to Tuk now are local residents and tourists—which, we noticed, were few and far between.

Several kilometers outside of Tuktoyaktuk, we pulled into a spot where you can hop onto a platform and enjoy a sweeping view of the pingos. Great idea, I thought to myself, until I got a whiff of something putrid. The source, we soon learned, was a sprawling garbage dump that sits on top of a wetland just meters from the Arctic Ocean. This is where all of Tuk's waste has been dumped for decades.

It was difficult to fathom how there was $300 million for a highway but not enough money to deal with this festering mess. The dump is just three feet above sea level, which will put it about a foot and half feet below sea level by the end of the century.

The town of Tuk faces the same fate. As sea levels rise, and as rapidly melting sea ice loses its ability to buffer the fragile, peat-covered shorelines from warmish, salty waves and increasingly powerful storm surges, the coastal town is slipping into the sea. Several meters of shoreline is lost some summers.

Residents of Tuk were fortunate not be dead center in a storm surge that swept seawater up to twenty miles inland into the Mackenzie Delta in 1999. Lakes and streams were inundated with seawater. When scientist Michael Pisaric and others were informed by the Inuit of what had happened, they came in to assess the impact. Shrubs were dead within a year of the surge, they concluded. An additional 37 percent of them died over the next five years. Ten years after the flood, the researchers found a striking shift from fresh- to saltwater species.

As we entered Tuk and passed the cemetery, now partially surrounded by a wall of riprap, I realized that the history of this community is increasingly being defined by what is no longer there. The Hudson's Bay Company building, which was relocated here when the post at Kitigaaryuit was no longer viable, washed away into the sea many years ago. Several other buildings have suffered the same fate.

There are many places in the western Arctic such as Crumbling Point that speak to the fragility of these rapidly thawing shorelines. Kitigaaryuit, now a national historic site, is sinking into the Mackenzie River and taking its graves and sod houses with it. The peaty shores of Herschel Island, a contender to become a United Nations World Heritage Site, is eroding at a rate of a meter a day in summer—six times the average for the previous 65 years.

None of this should come as a surprise. I was in Tuk in 2007 when Steve Solomon, a coastal geologist working for the Geological Survey of Canada, came to tell the Inuit what they could expect in the future. He and his colleagues had created a digital elevation model that simulated what would happen to the town of Tuktoyaktuk in 2050 if a major storm, such as the powerful one that buffeted the coast in 1993, swept in. Winds from that storm in 1993 were clocked at 62 miles per hour.

In his presentation, Solomon described how storm surges associated with that gale in 2050 would be about seven times the normal tide range of one foot. A storm like that would overwhelm many parts of Tuk. It would sever access to the airport in five different places and likely destroy the community's supply of freshwater. Homes would be lost. People would be stranded.

The residents of Tuk have only two realistic options. They can try to persuade the federal and territorial governments to invest heavily in a sea wall. Or they can ask for funding to relocate to higher ground, as the Alaskan community of Shishmaref has opted to do. If that were to happen, which seems likely, columnist Peter Foster will be proven correct. The Inuvik-to-Tuk highway will be a "road to nowhere."

The likelihood of the energy industry making another play in this part of the world seems dim following nearly a half century of failure. If it did, the road to Tuk would likely be of little use. In the fall of 2020, many parts of the road were flush with the surrounding environment. The gravel fill had migrated onto the neighboring tundra, giving one the impression of a flattened pancake," as geological engineer Katrina Nokleby, a member of the Legislative Assembly of the Northwest Territories, observed. "The ponded water along the roadside," she added, "is the kiss of death for permafrost."[16]

But the legacy of exploration continues, as I discovered when I was alerted to a civil case that quietly found its way to court in 2019. It involved exploratory wells that had been drilled into the Mackenzie Delta in the early 1970s by Gulf Oil. Gulf suspended the operation in 1973 after getting government approval to fill the top 2,000 feet of casing with diesel fuel. When the company abandoned the site in the mid 1980s, the diesel fuel was, according to court documents, "no longer in place." In short, it had leaked out.

Seemingly unaware of this at the time, the government concluded that this and other sites that Gulf had abandoned had been satisfactorily reclaimed.

When Shell bought Gulf's properties in 1991, the company soon discovered that these and other wells and sumps were leaking. Rather than fixing the problem, the company took ConocoPhillips, which had taken over Gulf, to court to see who would be responsible for a cleanup in the event that the problem was discovered by someone in government responsible for oil and gas inspections.

J. B. Hanebury, Master in Chambers, summed it up succinctly. "This application puts a legal cast on the old game of hot potato," he wrote. "A well site and its associated sump pit located in the Mackenzie Delta have been leaking contaminants for decades. No company has stepped forward to remediate the contamination. Instead, two companies each point the finger at the other as the owner responsible for the cleanup.

This action was started in 2014 and now, five years later, remediation has still not occurred, and the Court is asked to determine ownership of both the contaminated site and other well sites."[17]

This dispute underscored once again how the Arctic continues to be a wasteland for anyone other than the people who live there. The case, however, paled in comparison to what Panarctic Oils Ltd. did in the Canadian Arctic in the 1960s and 1970s. During its decades of operation, it drilled 150 wells, 112 of which were onshore. More than 11,000 miles of seismic lines were carved out on the tundra.[18] Panarctic's method of getting rid of tons of garbage—from junk steel and waste oil to a half-ton truck—was to drill a hole in the ice and use the Arctic Ocean as a dump. The company got caught, but only because an employee (the son of an environmentalist) told authorities.

No one in the company was punished when numerous efforts were made to frustrate the investigation.[19] What the Canadian government did while the investigation was underway was to help Panarctic find a way of dumping waste oil on the tundra on the High Arctic island of Ellef Ringnes. The geologist behind the experiment was a university permafrost scientist. He concluded that surface disposal on those polar, semidesert environments "where plant and animal productivity is low" may be the answer to get ridding of waste oil.[20]

When I got a chance to visit the island several years ago, there were not a lot of plants and animals to see. But there were expansive spreads of orangey-brown bryophytes that lit up the tundra the way the late-summer sun did when we drove the highway to Tuk. The one thing that struck me was how much evidence there was of that oil and gas activity. Deep, muddy depressions left behind by abandoned well pads, the tracks of trucks and bulldozers imprinted in the tundra, empty fuel drums, and the remnants of Weatherhavens were still there, just as they are on many other Arctic islands. The other thing that came to mind is that, since oil was struck in Norman Wells in the Northwest Territories in 1920, not a drop of oil from the Canadian Arctic, save for one shipload that was launched mainly for publicity purposes, has ever made it to southern markets.

Chapter 11

Tundra Beavers, Saltwater Trout, and Barren-Ground Grizzly Bears

We need the tonic of wildness—to wade sometimes in marshes where the bittern and the meadow-hen lurk, and hear the booming of the snipe; to smell the whispering sedge where only some wilder and more solitary fowl builds her nest, and the mink crawls with its belly close to the ground.

— Henry David Thoreau, *Walden* (1854)[1]

From the Mackenzie Delta to the British Mountains that border the Arctic National Wildlife Refuge in Alaska, much of the coastal landscape is impossibly low and dotted with thousands of lakes. This land is short on trees, but it is big on shrubs, sedges, and grasses growing out of peat that is soft and spongy for a few weeks in summer and rock-hard in winter. No one knows how much peat there is except that it is well represented, thawing and rapidly slipping into the sea along the coastal edges as Arctic ice disappears and as rising sea levels and storm surges relentlessly extend their reach inland.

Rivers like the enormous Mackenzie and the smaller Big Fish, Blow, and Babbage wind their way through these peatlands to the Beaufort Sea, where their fresh water is caught up and swirled around in the Arctic's most significant gyre. Most of these rivers flow gently. But the green, crystal-clear Firth that squeezes out of the high country in Alaska into the Yukon does it with big waves, deep gurgling holes, precarious ledges, dangerous drops, and heart-stopping waterfalls.

Freshwater thermal springs keep water open year-round at places such as the Aufeis on the Firth, and Fish Hole on the upper reaches of the Babbage watershed. Fish Hole is culturally important because Babbage River Dolly Varden, which are genetically distinct from other trout populations in neighboring watersheds, are critical to small coastal Inuit fisheries at Herschel Island, Shingle Point, King Point, and coastal Alaska.[2]

More than 200,000 caribou spend the spring and summer here and in the Arctic National Wildlife Refuge to the west, giving birth and fattening up on lichen and sedges. Where there are caribou, there are wolves and bears. Barren-ground grizzlies that dwell in this part of the world are genetically the same as grizzlies elsewhere. But these bears are about half to two-thirds the size, which works in their favor because big bears would have a hard time finding enough food to get them through six or seven months of hibernation.

Being small doesn't make them less powerful, as I discovered while accompanying scientist Andrew Derocher one week in spring when he was capturing and collaring them with satellite tracking devices. On one of the helicopter flights, we homed in on a grizzly that had just killed a bull caribou. The bear was half the size and weight of its prey, but powerful enough to drag it a kilometer uphill to a cache site, and fearless enough to rise up on its hind legs and take a swat at the helicopter hovering above.

We spotted the bear in the Husky Lakes area, where small streams such as Siksik, Little Bear, and Trail Valley Creek meander through the permafrost. Saltwater tides and storm surges that pulse in from Liverpool Bay and the Beaufort Sea make the lakes they flow into brackish.

Some of the seventeen species of fish found in the Mackenzie River watershed got here more than 8,000 years ago when meltwater channels from the McConnell Glacier icefield dried up, stopping the flow of water from the Liard River to the Yukon River, and opening a flow of water from the Bell, Rat, and Porcupine Rivers westward into the Yukon. Peat-bog formation followed, leaving Dolly Varden in rivers north of the mountain divide to evolve in isolation.

The plethora of fish explain why beluga whales from Alaska and Russia dare to venture into the lakes. The whales feed primarily on Arctic cod. But they will also eat squid, char, herring, capelin, sand lance, and even lake trout, which have evolved and adapted to salt water here. Occasionally the white whales stay longer than they should. Sudden deep freezes, such as those that occurred in the autumns of 1989, 1996, and 2007, trapped more than 800 of them. As the ice-bound, open pools of water got smaller and smaller, there was eventually no place left for the whales to come up for air.

The extraordinary nature of this tundra has gotten even more fascinating in recent years as Pacific salmon have begun migrating into the Mackenzie River system in relatively large numbers. While chum salmon have been doing this for millennia, all five species of Pacific salmon have now been caught by local fishermen. A few have made it all the way to Great Bear Lake, which is an epic swim into potentially hostile waters dominated by giant carnivorous lake trout weighing as much as eighty pounds.

It was cloudy, calm, and cool the morning that scientist Philip Marsh and I flew into his Trail Valley Creek research camp. Looking out through the plexiglass bubble of the helicopter, I searched in vain for a bear that might be harvesting berries below or for caribou that might be feeding on carpets of lime-green lichen. What stood out was a crater the size of a football stadium that had recently opened up along the side of unnamed tundra lake. The so-called thaw slump was gray, muddy, and barren—a sharp contrast to the brilliant russet and gold of the surrounding tundra vegetation.

These landslides occur as warming temperatures rapidly thaw permafrost. Slumping across the Arctic is increasing just as fast as the peaty shorelines of the western Arctic and northern Siberia are sliding into the sea. Many are smaller than this one. Some are much bigger. The Batagaika Crater in the Yana River Basin of Siberia is nearly a mile long and more than 300 feet deep—superstitious Yakutians call it "the doorway to the underworld."

"We're seeing slumping along shorelines that can drain most of the water in a lake in just days and even hours," said Marsh as the pilot circled the site so that we could get a better look. "It's not surprising when you consider that as much as 80 percent of the ground here consists of frozen water. When that ice melts, the glue that holds the frozen ground together literally falls apart."

One of many permafrost slumps in the Arctic that are resulting in lakes draining and water chemistry changing. (Photo: Edward Struzik.)

If the abrupt thawing continues to accelerate as it has done farther south in the Old Crow Flats along the Yukon-Alaska border, the landscape will look very different in the decades to come. In the half century leading up to 2007, lake drainages in the Old Crow Flats, according to Canadian scientists Trevor Lantz and Kevin Turner, have become five times or more frequent in recent decades, accounting for the loss of about 15,000 acres of lake area, and a gain of more than 200 lakes.[3] This catastrophic reconfiguration of the landscape is a big concern for the Vuntut Gwichin people who hunt, trap, and fish in the Old Crow Flats. Not only does it threaten the supply of food, but it also destroys important memories, as Old Crow Chief Dana Tizzy Tramm noted after Alma Lake suddenly drained. "That was my elder Norma Casey's family's traditional lake," he told a group of us." "For generations, upon generations, they had occupied this area, and she was decimated as a person. She was in tears."

The rapid thawing of permafrost, however, also has implications for the climate. There are 1,400 gigatons of carbon stored in permafrost.[4] (A gigaton is one billion tons). That's almost twice as much as there is in the atmosphere. Most scientists had assumed that the thawing would be gradual. But scientists Merritt Turetsky and colleagues are suggesting that the abrupt and erratic nature of permafrost thawing could release carbon more quickly throughout the circumpolar world. It may not be the carbon bomb that some predict, but a slow burn that could double the warming from greenhouse gases.[5]

Rapid thawing could buckle roads like the Inuvik–Tuk highway, destabilize buildings, and uproot trees that grow on permafrost. These "drunken forests" of trees that end up leaning up against each other are becoming an increasingly common sight.

When I asked Marsh if he and his colleagues had issues with bears at his Trail Valley Creek research camp, he was happy to report they had not had a dangerous encounter. The only regular visitors were a fox that members of his research team named "Monty" and a damnation of beavers that were staking out a place on the tundra, where they are not typically known to build dams.

As the pilot circled the research site looking for a dry spot to land on, one arm of the camp below looked a little like a stick man. The shape was formed by narrow wooden boardwalks that connect weather stations, snow and rain gauges, a precipitation radar system, and instruments that determine how much carbon dioxide and methane is being absorbed by tundra plants and how much of those greenhouse gases is being emitted into the atmosphere. The boardwalks were laid down so that the scientists' boots don't disturb the thawing peat or skew the recordings. Everything is powered by solar and backup generators.

Just as we were about to land, I counted ten two-person sleeping tents lined up in front of bigger Weatherhavens that are used for cooking, eating, and socializing as well as storing food and equipment. The camp, I noticed, was surrounded by an electrified bear-proof fence. Before I hopped out of the helicopter, I made a mental note to myself, recalling what it was like to accidentally grab an electrified bear-proof gate, as I once did at the Daring Lake Research site, 45 miles north of the treeline, where similar tundra and permafrost studies have been ongoing since 1994. It felt as if my teeth were about to be dislodged.

There are many reasons why Marsh decided to set up shop here. The Husky Lakes are at the epicenter of the most rapidly warming region on Earth. Ice-rich permafrost is thawing, shrub cover is expanding, less snow and rain is falling, and those thinner layers of snow are melting weeks earlier. Streamflow is also changing in complex and unexpected ways.

The impacts of these changes are often subtle and sometimes catastrophic, as we saw when we flew over that crater. Where there used to be cranberries, blueberries, cloudberries, shrubs, sedges, and lichen, there was nothing left but mud, silt, and peat that was presumably blowing off all of the carbon that it has been storing for centuries. If this had happened in an urban area, it would have swallowed dozens of buildings.

The slump didn't seem big enough to cloud up the lake and change its chemistry in a significant way like others have. In the lake-rich tundra uplands of the Mackenzie Delta, where Marsh conducted research studies for many years, about one in ten lakes have been affected in ways that have altered water chemistry.[6] In one notable case filmed by scientist Steve Kokelj, water from an upland lake spilled into the Peel River watershed when the cliff the lake sat next to collapsed. The waterfall drained approximately eight million gallons of water in just two hours.[7] Permafrost-related ions such as calcium and sulphate, along with heavy metals such as mercury, were flushed downstream into the Mackenzie Delta.

Mercury is another big concern in areas where permafrost is thawing. There is a lot more of it stored in the Arctic's frozen peat than anyone previously thought. In 2018, Paul Schuster of the US Geological Survey and his colleagues estimated there are 793 gigagrams (874,133 tons) of mercury, the equivalent in volume of 23 Olympic-sized swimming pools. Much of the mercury is naturally occurring. The rest comes from coal-fired power plants and other industrial sources.[8]

Mercury does little harm when it is attached to organic compounds frozen in permafrost. But when thawing occurs, microbes that break down those organics can transform it into methylmercury, which can be extremely toxic. If between 30 and 99 percent of the Arctic's new-surface permafrost thaws by the end of the century, as some predict, much of this locked-up mercury could be swept into rivers and streams and end up in fish-inhabited lakes and the Arctic Ocean.

Minutes after we landed, we headed to the cook tent, where I poured myself a cup of grainy, percolated coffee. Marsh offered me some whipping cream that he had brought along as part of his low-carb, high-fat ketogenic diet—which, he says, keeps him healthy enough to keep canoeing, skiing, and biking in his senior years.

My first contact with Marsh had come years earlier, when he was working for Environment Canada, the federal department tasked with assessing and monitoring change in the Arctic. He and colleague Lance Lesack had just published a sobering report predicting that a third of the 45,000 lakes in the delta could disappear in the coming decades as a consequence of warmer winters, less snow, and less ice-damming along the Mackenzie River.

Marsh is confident that he has most of the tools necessary to produce the hydrological models that are needed to determine what this new Arctic will look like. He just didn't expect tundra beavers to figure into the equations when he laid out his blueprint for research.

Permafrost scientists are a rare breed. Part of that, no doubt, comes from the nature of the work they're engaged in. When not drilling holes in intense cold, they are drilling holes in stifling heat while fighting off biting flies and mosquitoes. This would drive an otherwise sane person crazy. In an effort to maintain their sanity, some permafrost scientists play games like "page watch," where the goal is to kill as many mosquitoes and blackflies as possible by slamming a field book closed in the most timely fashion. The victor is the one with the most insects flattened on a page.

Permafrost scientists are such gluttons for punishment that Peter Kershaw, a University of Alberta scientist who devoted his entire career to the study, had the dubious honor many years ago of being named to *Popular Science* magazine's "Ten Worst Jobs in Science" list. He finished just ahead of the Orangutang pee-collector in the Borneo peatlands and two places behind the human-semen washer in Los Angeles.

Marsh entered the world of cold-climate hydrology in direct fashion, graduating from McMaster University, where one of North America's first muskeg laboratories was established in the 1950s. Scientists there became North American leaders in the field when peat was viewed more as a challenge for resource, road, and hydro development than for all of its many virtues. During those years, the National Research Council of Canada hosted annual conferences aimed at solving the "muskeg problem."

Marsh's PhD supervisor was Ming-ko (Hok) Woo, a giant in cold-climate hydrology. Woo supervised several brilliant hydrologists, including the University of Waterloo's Jonathan Price, York University's Kathy Young, Nigel Roulet at McGill, and McMaster University's Sean Carey. Woo and Mike Waddington were among the first to examine the impact that beavers have on the sub-Arctic forested peatland ecosystem in the Hudson Bay Lowlands.

Following graduation, Marsh was hired as a research scientist at the National Hydrology Research Centre in Saskatoon, working alongside glaciologists like Mike Demuth, whom I know well, having accompanied him several times into the Brintnell Icefields in the Mackenzie Mountains. While at McMaster, and during his time at the Environment Canada research center, Marsh worked on a variety of hydrological issues in places as far-ranging as Lake Winnipeg in Manitoba and Eidsbotn Fiord on Ellesmere Island in the High Arctic. His decision to leave the government job and join the faculty at Wilfred Laurier University in 2013 was fortuitous because it opened the door to more avenues of stable research funding and closer access to students who might join his research team.

Thomas Misztela was one of Marsh's MSc graduate students. When I met him, he had a boy's face, long hair, a rainbow-colored shirt, and a turquoise baseball cap that accentuated the fact that he was more of a free-wheeling spirit than the ten other researchers in camp. Not that this was a serious, stuffed-shirt bunch of young scientists. They liked to have their fun, as I learned later that first night when I was invited to drink a "polar molar"—an initiation ritual for newcomers like me.

The drink is modeled after the Sourtoe cocktail in the Klondike. Whereas the Sourtoe cocktail has a mummified human toe as the key ingredient in the alcoholic mix, the polar molar has a rotting tooth. This one was extracted from a caribou, presumably killed by wolves or a grizzly bear in the area. Seeing the sorry state of the tooth, I was initially inclined to pass up the offer. But I had a change of heart when Branden Walker, the man who manages the camp and engineers some technological wizardry to monitor ecosystem change, offered up a ten-year-old bottle of Scotch as part of the mix.

Misztela didn't stand out when he signed up for one of Marsh's undergraduate courses. He did make an impression a few years later, though, when he was out of school, working as a waiter and serving Marsh at a local restaurant. There was something about Misztela that night that convinced Marsh to take him on when he expressed interest in doing graduate work. It was the kind of hunch that many scientists I know are beginning to rely on more and more after realizing that high grade-point averages don't always augur success so well in challenging field-work environments.

When I emerged the following morning, Misztela and PhD student Evan Wilcox were in the cook tent gearing up to do some field work. Niels Weiss, a Dutch scientist who was a postdoctoral fellow at the Laurier University research office in Yellowknife, was slabbing cream cheese an inch and half thick on a toasted bagel while giving me advice on peatland ecologists I might talk to in the Netherlands.

Weiss's plan was to join Wilcox, who was helicoptering out to several lakes that he was studying. Misztela was staying put to retrieve a water gauge that got buried when a beaver built a dam on top of it. Intrigued by the prospect of seeing one of these tundra beavers, I asked if I could come along.

The morning was brisk, and the cold metal of the shotgun I was carrying numbed my fingertips. It's the protocol here to carry a firearm in the event a bear comes along and pepper spray and shouting fail to deter the animal.

"You can tell where the beavers are when you come to places where there is a clearing in the shrubs," Misztela said as we began bushwhacking through five-foot-tall alders and willows along the creek bank. "I saw my first beaver when I was taking a dam apart. That was a surprise."

"You ever see that video of the Russian guy who was taking pictures of a beaver before it bit him on the leg?" he asked me. "It tore through the artery and the guy bled to death. That's what I was thinking about when the beaver came out of its house and started slapping its tail, warning me to get out of there. I don't know if there's a difference between Russian beavers and Canadian beavers. But I didn't want to stick around and find out."

Trail Valley Creek flows from the northern edge of the boreal forest through the tundra. The permafrost here is more than 800 feet deep. Marsh does not yet know how deep the peat is, because he has not yet done an extensive coring program. But the active-layer portion of the peat, the upper layer that typically thaws, is ten to twenty inches deep.

I was told that one can normally jump across the creek at this time of year, when the melting snow is long gone and the around-the-clock sunlight evaporates what little is left. But the beavers were busy. There were at least seven dams blocking the flow of water on the creek. Misztela was wearing a life preserver as he hacked away at the dam with an axe, searching for the lost gauge. I instinctively drew back from the shoreline when, concerned that I might end up like that poor Russian, I heard the distinct sound of a beaver slapping its tail.

Misztela was dripping sweat when he finally chopped deep enough into the dam to retrieve the gauge. Out of breath, he removed the life preserver and sat down on shore, quenching his thirst with a long drink of water. Looking relieved, he asked whether I was interested in coming along to find another gauge that had gone missing.

"Seventy to eighty percent of the water that flows through here comes during the spring freshet," he said when we stopped for a while to pick through a particularly lush blueberry patch that had a sprinkling of cloudberries. "That used to begin the first week of June, sometimes at the end of May. This year the freshet started on May 21. Last year it began on May 13. Winters aren't as cold as they once were. I looked at the meteorological records over the past 50 years. The last time the temperatures dropped to minus-50 was in 1977. There's been a two- to three-degree increase in temperature over that time."

A longer growing season has clearly favored shrubs over lichen, grass, and sedges all across the Arctic, especially here in the western Arctic where the so-called greening is accelerating more rapidly than any other place in the polar world.[9] Making sense of what this shrubification will do to permafrost isn't easy, as scientists such as David Hik and Ryan Danby have told me many times during the course of their studies in the Yukon and Northwest Terrirtories.

Scientist Branden Walker records weather data at the Trail Valley Creek research site in the western Arctic. (Photo: Edward Struzik.)

Evan Wilcox did his best to explain the matter later that night, after dinner.

Shrub branches are good at trapping blowing snow and insulating the permafrost below. But the taller the shrubs grow, the more the dark branches radiate heat and melt the snow. Warmer soils can also trigger microbial activity in ways that offer up more nutrients for these deciduous shrubs to absorb. It's similar to the "ice-cold hot spots" of microbial activity that one sees in ice-filled depressions on glacial surfaces.[10]

This greening of the Arctic may be good for barren-ground grizzlies as well as southern plants and animals that are expanding their range into the region. But it's a challenge for caribou. The shrubs shade out not only the lichens that get them through the winter, but also the sedges and grasses that Arctic tundra specialists like caribou thrive on.

As Misztela was studying his GPS to give us a more exact idea of where the second missing water gauge might be, Barun Majumder joined us. Majumder was one of the other unorthodox decisions Marsh made when he was recruiting people to work for him. Born and raised in India, and schooled in the phenomenological aspects of quantum gravity, Majumder didn't strike me as someone who would be interested in spending a numbingly cold spring sleeping in a nylon tent in order to understand how snow cover forms and melts, and how the answer to that question might be used to better understand climate change and its impact on natural-resource development in this watershed.

His background in physics seemed to back my hunch. The title of one of his research papers was "I-Love-Q Relations for Neutron Stars in Dynamical Chern Simons Gravity." But when Majumder described to me the serenity of being out here when it is cold and eerily silent, and when the northern lights are pulsing across a moonless sky, I began to see why Marsh was so big on him. "There is nothing like this where I come from in India," he told me. "It is a magical place. I can think clearly here."

Majumder was calm and thoughtful in a way that reminded me of Winnie the Pooh. "Think, Think, Think." It's what I imagined he was doing as we followed the GPS. Some of what he told me then and later was beyond my comprehension. His wind models for drift formations-formations[11] begin with equations that start with

$$v = (h/L)\,f'(x/L)\,u_*\left\{\frac{\ln(\Delta y/y_0)}{\kappa} + \varepsilon\hat{u}\right\} + \frac{\varepsilon l u}{L}\,\hat{v}(x/L,\Delta y/L)$$

and end with numerics that are just as baffling.

What I did come to understand is that the persistent Arctic winds redistribute a lot of snow in winter, and that a lot of that snow ends up in the lower elevations of this valley.

"How much?" I asked him.

"Two and a half meters high where we are standing," he said. "Up to six meters elsewhere."

I assumed that most of it eventually melts into Trail Valley Creek. But Majumder wagged his finger and told me I was mistaken. Forty percent of it changes into water vapor before the spring thaw comes into play. "It sublimates during blizzards," he explained.

This is important, because the less snow there is in winter, the less insulation there is to protect the roots of plants from extreme thaw-freeze events that are becoming more common in the Arctic. And the less snow that melts in spring, the less moisture there is for the peat to absorb and store throughout the summer months. Longer, warmer, drier summers may eventually sap this moisture from the peat and further stress tundra plants like blueberries, which need at least 50 percent moisture in the soil to produce a good crop of fruit. One result could be an earlier browning of the Arctic when caribou and grizzly bears need to fatten up.

As we headed back to camp, the helicopter that flew Wilcox out to survey some of the tundra lakes was just returning. Over a lunch of peanut butter and jam on stale bread, Wilcox explained why it was important to figure out the role that precipitation plays in recharging the lakes in the region. Theoretically, less snow and rain, along with warmer summers and winters, will lower lake levels and cause streams like Trail Valley Creek to flow intermittently. This could have an impact on spawning fish, birds nesting in wetlands, and those adventurous beavers that are staking out new territory on the tundra.

Figuring it out has been challenging, though, because there are thousands of lakes covering half the tundra between Inuvik and Tuk. By the time I arrived, Wilcox had taken samples from 120 lakes across nearly 800 square miles of the area.

Even with all that data, he allowed that it was too soon for him to say whether there was a drying trend here, as there has been in the Old Crow Flats. The one body of water that had him scratching his head was one that appeared to have suddenly drained—not entirely, but enough to have dropped by several feet in the past few days. Wilcox was pretty sure it wasn't a landslide or thaw slump that was responsible. "It looks like a beaver dam collapsed," he said. "Maybe it was a heavy rainfall that caused the dam to burst. Or maybe the beaver just abandoned it. There was evidence of another beaver downstream. It could have been the same one."

As Wilcox and I chatted again later that night, the others were on their laptops trying to take advantage of a satellite internet service that allows researchers to communicate with family, friends, and colleagues and to keep up with news of the world. It's often hopeless because the signal is slow and spotty. "I liked it better when we didn't have it," Wilcox said. "It forced us to spend the evenings talking to each other. I also like the idea of not knowing what's going on in the outside world. When we finished the field season in the summer of 2015, we learned two remarkable things. The Chicago Black Hawks had won the Stanley Cup that June. And Donald Trump had announced that he was running to become president of the United States."

No one in camp was unaware of the fact that much of this land they are studying belongs to the Inuvialuit (Inuit of the western Arctic), who were granted legal claim to more than 35,000 square miles in 1984. If scientists want to conduct research here, they have to get permission from the Inuvialuit authorities. And if they want to conduct research over the long term, they have to explain how the Inuvialuit will benefit from their findings.

I recall the signing of that land-claims agreement well, because I was living in the Northwest Territories at the time. It's fair to say that not even the Inuvialuit were talking much about climate change back then. Winters were much colder than they are now. Fog was not a regular summertime visitor. The runaway growth of shrubs was not evident and there was still a lot of shore-fast ice to buffer the peaty coastlines from storm surges.

But then the warning signs, which were always there, began to present themselves in ways that could no longer be ignored. Water levels in shallow upland lakes in the delta started to drop. Pooling of water began to increase in other areas. Summer storms, such as the ones in 1993 and 2000, swallowed up as much as 23 feet of shoreline around Tuktoyaktuk. Some of those storms triggered surges that extended their reach 20 twenty miles inland. Caribou numbers plummeted. Grizzly bears were mating with polar bears. And mercury, which is stored in peat and permafrost, started showing up in beluga whales in alarming levels.

As sea levels rose and permafrost thawed, Kitigaaryuk, once the site of the largest gathering of Inuit in the Canadian Arctic, began to sink, and the 190 graves, 17 houses, and the foundation of a Hudson's Bay Company building began sliding into the sea. When archeologists and permafrost specialists were brought in to see if the area could be shored up, they concluded that it would be prohibitively expensive.

The 800 people living in Tuktoyaktuk are facing a similar threat. It's becoming increasingly clear that no amount of dredging or piling of riprap is going to save the town from rising sea levels and future storm surges. There is no bedrock on the coast—only ice-rich permafrost. That ice is eventually going to thaw.

On the morning of my departure, I got up early to see the sun rise on the tundra, and to pick some blueberries and cranberries that were in good supply around the research station. I did not need to worry about the electric gate because it was not humming.

It was dead calm when I heard what is normally a springtime call in the heaths, moors, and bogs of the tundra world—the nasal cluck of a willow ptarmigan. I laughed because there are few sounds in nature more comical than this. Maybe he was still looking for a female to mate with. Or maybe he was just letting me know who was boss out here.

Just as I was about turn back to the camp to pack up my gear, I was alerted by the slap of a beaver, issuing a warning to the presence, maybe of a wolf or a bear that was nearby. The old Arctic was disappearing and a new one was rapidly unfolding.

Chapter 12

Portals to the Otherworld

Sweet is the swamp with its secrets,
Until we meet a snake;
'Tis then we sigh for houses,
And our departure take.[1]
— Emily Dickinson

In the summer of 1961, geologist John Fyles was hiking along the icy shores of Strathcona Fiord in Canada's High Arctic when he saw a patch of dark peat protruding from a hilltop in the distance. Peat in the polar desert region is of particular interest to geologists because, under certain conditions, peat can point to seams of coal, both anthracite and lignite.

When Fyles climbed up to have a closer look, he found branches of trees embedded in that thick layer of frozen peat. This was a remarkable find in an arid tundra landscape that is mostly covered in ice. The nearest living tree was located more than a thousand miles to the south.

The branches turned out to be 4.5-million-year-old relics of an ancient fen where peat nearly eight feet thick piled up in places. What made this find at Strathcona Fiord even more extraordinary were the bite marks embedded in the wood fiber. When paleontologist Dick Harington got a chance to look at the specimens many years later, he recognized them to be the dental imprints of a beaver, another specimen that does not currently have a home anywhere in the Arctic archipelago. Those beavers, subsequent excavations revealed, shared the fenlands with three-toed horses, ancestral bears, and other animals that lived in the Arctic when it was much warmer than it is now.

Peat and the acidic bogs and fens in which it accumulates have long been associated with the "otherworld" where gods and spirits dwelled before rising up at night to terrorize local inhabitants. There's as much science to it as there is myth, because the anaerobic nature of peat preserves the past in detail equal to or greater than that delineated from ice cores, diatoms from lake bottoms, or, for that matter, stories told from one generation to the next. Peat is a medium that does not embellish the truth as people sometimes do.

Degrading or destroying peat for coal mines and other purposes, such as the one that was planned for Strathcona Fiord several years ago, is akin to destroying an archival record of climate change, pollution, and species diversity. The stories preserved in peat and in permafrost tell us how much lead the Romans put in the atmosphere, how much of the mercury in the oil sand peatlands comes from industrial activity, and how the chemistry of soils has been impacted by volcanos and ancient wildfires. (Scientist William Shotyk was heartened to discover that there are still some bogs in the oil sands regions that are exceptionally pristine.)

What the record of peat also tells us about the past is that the Arctic was a lot warmer for most of the past 100 million years than it has been since the Last Ice Age, and that woolly mammoths and other animals may have hung on longer than previously thought, as permafrost scientist Duane Froese and his colleagues have recently gleaned from genetic material that was frozen in peat and permafrost in the western Arctic. He calls them "ghost herds."

This was not the first time that Harington found evidence of ancient beavers roaming the Arctic. He unearthed the remains of giant beavers (*Castoroides ohioensis*) along the peaty banks of the Porcupine and Whitefish Rivers in the Yukon. These 220-pound rodents were the size of a modern-day black bear. Instead of consuming fiber from trees, as much smaller modern-day beavers do, and as those little ones at Strathcona Fiord did, these behemoths fed primarily on aquatic plants.[2]

The exquisite condition of fossils such as these is easily explained

Cotton-grass meadow in a peatland at Strathcona Fiord on Ellesmere Island in the High Arctic. The fossils in a 4.5-million-year-old beaver pond nearby offer an intriguing look at what kind of plants and animals lived there back then, when the Arctic was much warmer. (Photo: Edward Struzik.)

by the fact that peat is an excellent natural preservative. Eighteen-hundred-year-old kegs of butter, a 1,200-year-old book of psalms, a 2,000-year-old fully clothed body, and a 10,000-year-old woolly mammoth, complete with pink flesh and fur, are among many specimens that have been recovered from peat bogs from around the world.

Cold temperatures and mosses such as sphagnum are a key part of the preservation. When sphagnum moss takes up nutrients such as potassium and magnesium, it gives up hydrogen ions, which acidifies the water. When sphagnum dies, it does so in a way that neutralizes nutrients that acid-intolerant plants need in order to grow. Starved of this plant energy, bacteria that may be present cannot easily engage in the composting that typically takes place. It's the reason why the Vikings and many other seafarers preferred to carry bog water rather than water from streams and lakes on their long journeys. Bacteria, viruses, and parasites such as giardia do not proliferate as easily in bog water as they do in less acidic ponds and streams.

This explains how the hoard of swords, shields, and trumpets that were dropped into Llyn Cerrig Bach, a small lake in Wales, in order to appease the gods of the otherworld were found to be in such good condition when workers digging peat found them during the Second World War.

The most ghoulish example of this kind of preservation is in the so-called bog bodies that have been unearthed in peat bogs in Scandinavia, the Netherlands, Great Britain, Ireland, and elsewhere in northern Europe. One of the best preserved was Tollund Man. He was found in 1950 by two Danish brothers harvesting peat for fuel from a bog near the town of Bjældskovdal. Seeing how the eyelashes, whisker stubble, and face wrinkles were still intact, they concluded that the death was relatively recent, and that the body may have belonged to a young man who had recently gone missing. A noose wound tightly around his neck suggested he may have been murdered. Police were called in to investigate. When they saw no evidence of digging in and around the peat, they called in forensic experts associated with the local museum.

The experts determined that the man lived for forty years and died in winter sometime during the Bronze Age (3200–600 BC). It was unlikely that he was killed by enemies, because someone took the time to carry him to the bog and close his eyes and mouth after placing him in a sleeping position. The educated guess is that he was a human sacrifice.

In 2011, an even older body was found by an Irish farmer milling for peat at Cúl na Móna bog. Cashel Man, as he has come to be known, was between 20 and 25 years old. CT scans revealed that one of his arms was broken and that an axe was used to strike him in the back several times. Archaeologist Eamonn Kelly believes that the nature of the injuries is evidence that the practice of sacrificing young men—a ritual connected with kingship and sovereignty well known from the Iron Age—was much older than previously thought.[3]

When I got the opportunity to join Harington in the field many years ago, the first thing he did was take to me to the beaver pond site to get the lay of the land that we were going to excavate the next day. The fossil-bearing stratum consisted of a two- to four-meter-thick sequence of mossy peat with various layers of sandy bedding, some cobbles and boulders, and larch tree trunks up to nine feet long.

It was surreal standing there on that peaty hilltop. The coming of summer, it seemed, had stalled on its march north. The surrounding mountainsides were dusted with a fresh coat of snow, and a brisk northeast wind was sweeping more of it toward the Prince of Wales Icecap to the east of us, looking like dust devils I had seen in the Mojave desert. Raging meltwater that flowed out of that icefield into the Taggart River failed to make much of a dent in the sea ice that covered the fiord as far as the eye could see.

Harington described to me how, at the end of that first day of that first field season at Strathcona Fiord, he and his companions were getting set to leave for camp without anything to show for their work when John Tener, a friend and fellow scientist, noticed something in a chunk of peat near the edge of the site.

It was like a concretion. When Tener opened it up carefully, it came apart like a peach. When he looked inside, he realized he had come across something really important. He immediately passed it over to Harington.

Harington told me that he didn't see what it was at first. Only after he turned it to the sunlight did he notice the vertebrae, ribs, and skull parts of what turned out to be a fish, a new genus and species of perch. "It was," he said, "really quite exquisite."

Following that first summer, Harington and Tener returned to the site six times with various colleagues and assistants: Clayton Kennedy, a senior technician with the museum, Natalia Rybczynski, a young scientist who eventually succeeded Harington when he retired from the Museum of Nature, Kim Aaris-Sørensen, a vertebrate paleontologist from Copenhagen, and David Gill, an experienced Arctic field worker. During one particularly productive field season, they recorded more than 150 significant finds. Among them were the teeth and partial skull of a horse that would have migrated to North America by way of the Bering Land Bridge three to four million years ago.[4]

One lesson that paleontologists, paleobotanists, and peatland ecologists learn quickly is that patience is a virtue. Sifting through the peat that we shoveled hourly into pails was monotonous if not grueling work. Each morning when I was there, we got up and climbed that small mountain. Hunched over, with eyes focused on the brown, frozen peat, we spent hours on end searching for anything that might look like it didn't belong. Initially, to untrained eyes like mine, it was maddening—not only because I did not have a clear idea of what I was looking for, but also because I had a hard time making that leap of faith to believing there was anything of importance to be found.

A few sticks, several snails, and a piece of enamel from an unknown animal were all that came my way those first few days. The gee-whiz moment arrived when I found a branch that looked like it had been chewed on. When I passed it over to Harington, he confirmed my hunch. The marks made by the bite of a beaver.

On the last day of that scientific expedition, Harington took me back to the outer edges of the beaver pond to describe how this all came to be when the glaciers and icefields began to grow and spread and take over the site. It was instructive because it brought home the fact that nature on its own does not stand still, even if we are not hurrying it along by degrading so much of it.

"You need only to look at this stark landscape all around us to realize what a different place this must have been four and a half million years ago, when the beavers and other animals were here," Harington said. "It is easy to imagine the massive snout of the glacier that left these high mounds of rubble moraine just before it melted back some 8,000 years ago. Clearly marked, raised beaches on its margins tell of the rapid rise of land during the last few thousand years. It is remarkable when you consider how catastrophic the changes were when the glaciers enveloped this part of the world."

The evening before my departure day, the sun finally broke through that impenetrable veil of fog and snow that had made our excavations so tiring. It was shortly after midnight. Unable to sleep with the sun shining brilliantly through my yellow tent, I decided to take a short hike along the sandy coast to the mouth of the Taggart River. Upstream on my side of the river was a verdant meadow of cotton grass growing out of peat that was as thick and spongy as anything I had seen in all of the peatlands I had explored. It was enchanting because there was nothing but rocks and boulders on the other side of the river.

Looking down just then, I found the fresh prints of a wolf. Through my binoculars, I could see no sign of the animal. What I did see was a patch of brown that turned out to be the boot that had gone missing after I slipped it into the vestibule of my tent the first night. How it had vanished had puzzled me to no end because Harington and Tener were too serious and seasoned to be practical jokers. Judging by the chew marks, the wolf had tried to make a meal of it.

Several months after I returned for home, I got a call from another one of the scientists who was conducting research at Strathcona Fiord, informing me that a mining company had staked a claim there in order to mine the area for coal. The company withdrew a year later when the Inuit of the region objected. The last time I looked, the scheme was still on the company's website, perhaps waiting for another opportunity in the future.

Chapter 13

"Growing Peat"

Bogland is an obstructive, argumentative, quibbling, contentious terrain; it demands step-by-step negotiations.
—Tim Robinson, *Connemara: Listening to the Wind*[1]

In 1991, the Johnson and Johnson company introduced a peat-based sanitary pad that promised to be thinner, lighter, and more environmentally friendly. More than $40 million was spent over a fifteen-year period to develop it. They spared no expense promoting it in major daily newspapers like the *New York Times* and the *Wall Street Journal.*

Company officials described it as being made from "a honey-colored plant" (sphagnum moss) that is "grown in the cold, clear waters of Canada." The moss, they said, was once used by native American women in its raw form as sanitary protection and for diapering babies. It was also harvested during the First World War, when sepsis, a potentially lethal reaction to infection, threatened to exhaust the supply of cotton used for bandages and uniforms. Medical experts tried everything from treating the wounds with chlorine solutions to making bandages out of carbolic acid, formaldehyde, mercury chloride, and cotton. They settled on sphagnum, which Irish soldiers used to treat wounds during the Battle of Clontarf in 1014.

This wasn't the first time a company saw profit in peat that might serve the hygienic needs of nonnative women. When the First World War ended, a company based in Portland, Oregon, created a napkin called Sfag-Na-Kins, which was manufactured with paper and sphagnum moss. The venture lasted only a year because of competition from cheaper, cellulose-based products.

The peat that Johnson and Johnson used for their product, sold in 85 countries, was extracted from a bog twice the size of Central Park in the Saint-Fabien-sur-Mer region of Quebec. The company, however, decided to stop producing the napkins in the late 1990s, partly out of concern that the environmental message it was promoting might backfire if women learned that the products they were using resulted in the draining of a wetland.

Faced with the need to restore the bog once the project ended, Johnson and Johnson looked to Dale Vitt, one of the first peatland ecologists in North America, to restore a peatland extraction site to an ecologically functioning state. A serendipitous moment put them in touch with Line Rochefort, a Canadian scientist who had done graduate work with Vitt before joining the faculty at Laval University in Quebec City. She was ideal because she spoke French and was much closer to the bog site.

Restoring peatlands is a promising way of storing terrestrial carbon, regulating climate, filtering water, mitigating floods, and protecting low-lying coastlines from rising sea levels and storm surges that are extending their reach inland. It may also be the best way for peat moss companies, already banned from extracting peat in many parts of Europe and the United Kingdom, and increasingly in some parts of North America, to stay in business.

The science of "growing" peat is a centuries-old work in progress that continues to be elusive, depending on the criteria used to measure success. There are those like Dutch peatland ecologist Hans Joosten who insist that restoring peatlands to their original functioning state cannot be done, and those like Rochefort who are convinced that success is already at hand and that near perfection may not be far off as scientists strive to better understand how the interactions among water, vegetation, microbial communities, and climate promote the growth of moss, the building block in peat production. For Joosten, the solution is to stop extracting peat altogether and rewet peatlands that have been degraded. For Rochefort, it's conservation as well as helping the peat industry achieve sustainability.

The idea of growing peat began sometime around 1658, when Martin Schoock, a professor in the Dutch city of Groningen, wrote the first book on peat. In one chapter, Schoock asks: "*An material cespitita effossa, progressu temporis restaurari possit?*" Roughly translated, this means: "Can this excavated combustible matter, after the course of time, be restored?"

It was an important question. Two-thirds of Central and Northern Europe was once covered in trees.[2] After most of those trees were felled, peat extracted from bogs and fens replaced wood as a primary source of energy. As time went by, many in the Low Countries, Germany, and elsewhere in the North Sea region feared that if peat extracted from fens and bogs were to be exhausted someday, there would no longer be a reliable source of energy, nor the revenue that came from the sale of peat. It was akin to concerns that arose during the energy crises of 1973 and 1979, when the shortage of oil and gas sparked fears of economic collapse.

Because the Dutch were so skilled at draining peatlands for energy and agricultural purposes, the country now known as the Netherlands was well on its way to becoming a superpower in the sixteenth and seventeenth centuries. It was wealth from peat that gave the Dutch a dominant navy, control of the silk and spice trade, colonies in North America, South America, Africa, and Asia, and the affluence to support renowned artists such as Hieronymus Bosch (c. 1450–1516), Pieter Brueghel (c. 1525–1569), Rembrandt van Rijn (1606–1669), Johannes Vermeer (1632–1675), and Jacobus Mancadan (c. 1602–1680). No other country as small has been so dominant in so many ways since the Dutch Golden Age ended with a series of disastrous events that followed the Franco-Dutch War in 1672.

For the Netherlands, the extraction of peat was often more valuable than the harvesting of grain. Wheat, oat, and barley did not grow well on peatlands that had been drained as had been done progressively since 1050, when people in Delft in South Holland began mining peat in the estuary of the Maas and Rhine rivers. The remaining peat eventually lost its buoyancy, sank, and compressed the underlying layers of peat below the water table.

Grain crops literally drowned in those wet conditions. (Cows did not, which is perhaps the reason why cheese and other dairy products became another measure of the country's considerable wealth.) By some estimates, peatlands were three times more valuable than farmland. So the Dutch continued to do what the residents of Delft taught them when they finished draining the bogs: they canalized the water from the polders and the mires, and they built cities on top of them.

This went on for the next three centuries. Municipal officials who attended a general meeting of cities in Holland in 1453 were so concerned about the possibility that the country's reservoirs of peat might one day disappear that they proposed banning the export of peat altogether. In 1514, fourteen cities in Holland informed tax officials that they derived all their income from this one resource.[3]

No one seriously listened, as Hans Joosten, water historian Petra J. E. M. van Dam, and others have documented so succinctly in several books and essays describing the demise of the *hoogvenen* and *laagvenen*, Dutch words for the raised bogs and low bogs that so dominated the country's landscape at one time. There was simply no alternative to peat.

By the seventeenth century, the peat in the bogs and fens around Amsterdam was almost exhausted, forcing miners into wet, sodden outbacks such as Groningen in the lowlands of the northeast, where scholar Martin Schoock penned his book on peat and where artist Jacobus Sibrandi Mancadan put his brush to one of the first paintings of peat being harvested. (German artist Albrecht Dürer is thought to have painted the first in 1495.) Holland in the seventeenth century was like Alberta in the 1960s, when conventional oil reserves were in decline and the provincial government was offering energy companies generous subsidies to find a way to extract bitumen from the oil sands in peatland country.

Enticed by similar subsidies and favorable taxation, Dutch peat-mining companies—*veencompagniën*—bought up huge swaths of land that were previously deemed to be worthless in order to produce energy and create more farmland. From that point on, it was no longer the business of turf farmers (*turfboeren*) harvesting peat for income in the off-season. It was a full-time, well-paying trade overseen by a complex corporate structure, and aided by engineers such as Cornelius Vermuyden, who designed and built dykes, reservoirs, sluices, channels, and windmill-generated pumps to drain the bogs and fens in a more systematic way. Vermuyden was so good at what he did that the English commissioned him to drain many of their bogs and fens.

Many in the Dutch peat industry, meanwhile, had come to the realization that sustainability plans had to be put into place. Extracting peat from one side of the bog or fen and allowing it to regrow on the other side was viewed as the best and most practical option, because most everything that was needed to grow peat—acidity, anaerobic conditions, seeds from donor plants, and cool, moist weather—were nearby, waiting for wind and water to transport them into the recovery site. This passive approach didn't work, though—at least not quickly enough to keep up with the rapid rate of drainage and extraction that came with mechanized mining.

The complexity of growing or accumulating peat quickly was so elusive that Robert Angus Smith, a nineteenth-century Scottish chemist who championed the idea of growing peat to produce energy in the British Isles, had doubts that it could be done. "I can quite imagine a peat-farm of no excessive size growing fuel as fast as it can be made use of by itself and neighbours, or if compression should succeed, growing oil and perhaps ammonia, when that can be economically removed," he wrote in an 1876 essay, "The Study of Peat." "But care would be required to have the proper plants and to feed them properly; and that is not, so far as I see, to be expected in the immediate future, without some previous cautious experimenting."[4]

The experiments never followed with any zeal, because charcoal, coal, kerosene, and whale oil (derived from massive slaughter of Arctic whales) eventually replaced peat as primary sources of energy. There were, however, notable exceptions in places like Ireland, Sweden, Finland, Estonia, and northern Russia, which lacked access to cheap coal. (Recognizing the importance of peat, the Swedish government appointed a Crown peat engineer in 1901 to survey the country and report on the quality of peat that might be used for fuel to drive locomotives and other industrial steam engines.)[5]

As interest in harvesting peat for energy waned when cheaper alternatives became available, entrepreneurs began exploiting peat for other uses. Peat was mined to produce bandages, litter for livestock, insulation for buildings, and horticultural soil conditioners. The peat industry was like the oil industry as the latter became creative in promoting cheap petrochemicals for the manufacture of clothing, cleaning products, food additives, and food coloring. As long as there was a cheap source of supply, there was someone there to drain the bogs and fens and dig up the peat.

By the end of the twentieth century, most of the expansive peatlands formed by the Rhine and Meuse rivers in Holland were gone. Eighty percent of the peatlands in the United Kingdom and the Republic of Ireland were badly degraded or done for. Nearly half (44 percent) of the peatlands that were left in Europe were no longer producing peat. If Russia is excluded, notes Joosten in his painstaking tally of the losses, a little more than half of Europe's peatlands can be considered dead.

When demand for peat eventually exceeded the ability of the United Kingdom, Holland, Germany, and other European countries to produce what was needed, entrepreneurs from Canada and United States stepped in. They did well until scientists like William Niering began making the case for wetlands as refuges for rare plants and birds and as filters for polluted water flowing into streams and rivers.

Niering, a professor at Connecticut College, was a visionary who was involved in the early development of The Nature Conservancy and their purchase of bogs in Pennsylvania and Connecticut long before anyone valued wetlands like those. When he wasn't tramping through those bogs and coastal marshes, taking notes on the back of a used envelope, he was waging war on the American lawn, encouraging people to raise their own chickens, and struggling to get his college to recycle. His wetlands research help lay the groundwork for passage of Connecticut's Tidal Wetlands Act of 1969.

Over time, other governments slowly bought into the wetland conservation imperative that came into play after the Ramsar Convention on Wetlands of International Importance Especially as Waterfowl Habitat, an international conservation treaty, was brought into force in 1975. But there was little in governmental decision-making in those early years that recognized the differences among bogs and fens and grassy marshes and forested swamps. It was such a gray area of study that wetland scientists were still debating how to classify wetland types in the 1990s.[6] A 1994 story in *National Geographic* underscored how indifferent the public was toward peatlands even as grand as the Everglades. "Maybe this land—rattlesnakes and all—is not so lovable: while our heads say ecosystem and biodiversity, our hearts still say swamp."[7]

As the value of peatlands came into sharper focus, people from all walks of life began to support the restoration of degraded landscapes such as the Great Fen in England and De Ronde Venen ("The Round Fens") in Holland. Supporters would soon include celebrities such as Prince Charles and actor Stephen Fry, who respectively became patron and president of the Great Fen restoration project sponsored by the Wildlife Trust for Bedfordshire, Cambridgeshire, and Northamptonshire.

Watching all this unfold, many in the peat industry in North America were justifiably concerned that the extraction of peat would come to an end, as it pretty much did in Holland by the early 1990s. Some were intrigued when heavily harvested peatlands such as the one that had been abandoned at Seba Beach in Central Alberta was restored with some degree of success by Dale Vitt and his research team.

The bog that Rochefort and others restored for Johnson and Johnson was among the first of more than a hundred peatlands she has worked on. When I first contacted Rochefort, she was getting set to go to Mongolia, where the country's peatlands were being dangerously degraded by permafrost thawing, drought, and overgrazing by yaks. A trip to the High Arctic to see how snow geese were affecting wetlands on Bylot Island would follow before the university year began. So she was busy that summer; she wasn't sure she had time for a meeting.

With a little persuasion, though, Rochefort agreed to take me on a tour of several peatland restoration sites that she oversaw, but only on the condition that I first visit a peat-extraction farm in Saint-Fabien in the Lower Saint-Lawrence / Gaspé region of Quebec so that I could better understand what she was doing. The farm is one of several in the region that produce peat for horticultural purposes.

I had never been to the region. The only image I had of it came from reading *Kamouraska*, Anne Hébert's haunting novel about a Madame Bovary–like woman who is a prisoner of a loveless marriage living in a nineteenth-century boreal world dominated by cruel religiosity, the white immensity of snow, and the misogyny of men. My good friend Jacques Després knew the place well enough. He had grown up in rural Quebec along the Saint Lawrence River before moving to New York to train at the Juilliard School to become a concert pianist. If there was one thing I should know about the Gaspésie and the Lower Saint-Lawrence, he told me, is that it is the cradle of the province, home to a proudly nationalistic group of people who have endured countless hardships over the centuries. What I would encounter would not be anything like the multicultural city of Montreal, where my wife Julia comes from.

I had just come off research trips to peatlands on the Hawaiian island of Kauai, a remnant fen in the Mojave Desert, the mountain fens of Colorado, the frozen tundra of the Arctic, and the wetlands around Georgian Bay. I was still trying to wrap my head around boreal sundews in a tropical peatland, polar bears denning in palsas (peat mounds with a permanently frozen core), pupfish in the desert, rich fens at 14,000 feet, beavers on the tundra, and rattlesnakes hibernating in layers of peat in the Great Lakes region. Were there really 546 plants in the boreal forest that indigenous people used for medicinal purposes?

What little I knew about the peat industry in Canada came from skimming through what little there is to know about its history. The first commercial peatland extraction operation in Canada was established in Victoriaville, Quebec, in 1864 to provide fuel for the Grand Trunk Railroad. When mixed with wood and coal, it produced a dirty, sputtering fire that shot out sparks and embers, igniting scores of wildfires along the route. Instead of giving up on turning peat into a source of fuel, the Canadian government spent a small fortune setting up an experimental station at the Alfred Bog in Ontario. After more than half a decade, it gave up in 1918 due to the First World War and the failure to turn Alfred Bog peat into reasonably clean energy.

The very serious business of extracting peat in the Lower Saint-Lawrence / Gaspé for agricultural purposes got its start in 1928, when

Francois Xavier Lambert founded a company that continues to bear his name. Ernest Meyer, a German Jew who moved to New York City to escape fascism, followed a decade later when he came to the region and bought the start-up Laurentide Peat Moss Company near Rivière-du-Loup, Quebec. As the son of an engineer who devoted his career to producing and selling peat in Europe, Meyer was well positioned to turn his new company, Premier Peat Moss Production, into one of Canada's largest peat producers.

Not everyone in Quebec was happy with these new business ventures. A local parish priest believed that a French Canadian *habitant* like Lambert had no business dealing with English entrepreneurs. The bogs, he insisted, would be better cleared of trees and drained to produce farmland. That was, in essence, the definition of *habitant*.

There was, however, no stopping this business from expanding. When Bernard Bélanger, a local who worked for Premier, took over from the Meyer family in 1978.

And then there was Huguette Théberge. She lived just a few miles down the road and looked on with envy on as these local companies were making modest, and in some cases very handsome, profits extracting peat from bogs and turning it into soil conditioner. Her husband Alcide Berger wasn't interested in quitting his job to help his wife until the small business began paying dividends.

In those early days, most of the harvesting at Berger Peat was done by hand and shovel, electric presses, tractor, and trucks. Bigger operations, like the one Bélanger took over, used vacuum trucks, rail cars, and miniature locomotives to transport the peat for export to the United States. When the Bergers were too old to continue the work, their son and nephew took over before turning the operation over to two granddaughters.

I had once worked on a tobacco farm where making a lot of money didn't make the farmers any more sophisticated. Given what Jacques had told me, I imagined being greeted suspiciously by two tough women, possibly with cigarettes hanging out of the side of their mouths while they spoke French in the laid-back Québécois twang that even the European French have difficulty understanding.

I have to admit that I was deeply embarrassed when two attractive women dressed in sharp business attire opened the door of a small prefab building and graciously invited me to sit at a table where smoked salmon, braised scallops, sushi rolls, cheeses, fresh fruit, and sparkling water were waiting for us. After a brief greeting, Valérie Berger excused herself to attend to other business.

It was left to Mélissa to tell me the story of how she and her older sister had ended up running this business.

"I was living in Manhattan, working for L'Oréal [the cosmetics company], when my father came to visit about ten or eleven years ago," she said. "He told me that he wanted me and my sister to take over the company. I wasn't really interested, because I had a good life and a good job in New York. It was when I came home for a visit that I saw the possibilities. And my father can be very persuasive."

It was obvious from the outset that this was more than just a family operation. The company has seventeen harvesting sites in the boreal peatlands here and in other parts of Quebec, New Brunswick, Manitoba, and Minnesota, and mixing plants in California and Texas. By the time the sisters took over, the company was already well into expansion, thinking, as other local companies were, about the day when they might find themselves with no more wilderness to move into because of conservation concerns.

It's tempting to compare the extraction of peat for horticultural purposes to the peeling away of peat to mine bitumen in Alberta. But there are differences in scale of degradation as well as important cultural differences to consider, as I found out when I was introduced to many of the employees at Berger and at Premier. Whereas people from other parts of Canada come to Fort McMurray for a short time to make their fortune in the oil sands before returning home, many of those working in the peat industry in Quebec are either third-, fourth-, or even fifth-generation locals. Young people who had left because there was no work returned when the peat operations were expanding and generating enough local revenue to sustain stores, restaurants, and entertainment centers. Investing in restoration, Mélissa told me, will ensure that their children will have a place to work as well.

Growing peat on land where the top layers have been extracted, I learned from Rochefort and from Sandrine Hugron, a member of

Rochefort's Peatland Ecology Research Group, is a lot easier than doing it in the oil sands. It begins with what she describes as "building the basement before slapping on bricks and wood." Mosses are the cement she uses to build the foundation of that basement. Sphagnum for bogs, brown mosses for fens. "Once you have the foundation for the basement in place," she said, "the rest will come"—reminding me of the famous line in the movie *Field of Dreams*.

This can happen naturally, as the Dutch understood three centuries ago, or it might not happen at all, as I discovered when Rochefort took me to the Bois-des-Bel bog seven miles northwest of Rivière-du-Loup. The harvested peatlands that were restored there, beginning in 1999 by Rochefort and Jonathan Price, were as starkly different from those that were left untouched for scientific analyses as the verdant, restored fens in the San Juan Mountains of Colorado are different from those that were degraded and then left untouched for a century.

As it accumulates, peat is a little like ice that grows in the Arctic. Young ice such as slush and frazil ice is very fragile. Wind, powerful currents, and warm upwellings can easily turn it back into water. Sea ice that survives the summer melt gets thicker and less salty as years go by, making it less vulnerable to those eroding forces. Like old ice, thicker, older layers of undisturbed peat can, after a time, withstand drought, wildfire, roads, and seismic lines so long as no more than two or three are at play, just as peatland ecologist Mike Waddington had told me.

The key to longevity is in the pore structure of old peat. According to University of Waterloo ecohydrologist Colin McCarter, the lead author of a paper on the subject, the older and deeper the peat, the greater the number of small pores there are to restrict moisture loss. Moisture is critically involved in the microbiological and biochemical processes involved in the breakdown and buildup of peat.

For peat to accumulate quickly, as Rochefort and others are overseeing, the harvested site is contoured to allow water and minerals to flow in. Berms and dikes keep some of that water from flowing out. Abundant, less desirable plant species like birch are then removed before donor plants such as sphagnum and brown mosses are introduced. After fragments of moss are scattered over the exposed peat surface, straw is used as a cover to keep the mosses moist. Small amounts of fertilizer such as granular rock phosphate are then added to stimulate the growth of *Polytrichum* mosses that help stabilize the ground surface; these are known to be good nurse plants for peat-forming mosses.

What comes up isn't perfect, as Rochefort readily admits. Plants such as the three-seeded sedge, mountain holly, cloudberry, and pink lady's slipper orchid remain absent at some of the restoration sites she showed me. But there were carnivorous sundews, pitcher plants, Labrador tea, and other peatland specialists. In one case, about 82 percent of the plant species had been restored in less than a decade.

"The other plants will come eventually," she said. "We're working on it."

Rochefort is irrepressible, a nonstop talker who chills out by skiing black-diamond runs whenever she is at home in her house at the bottom of a ski hill (her choice). As far as I could tell, there is no ebb in the flow of her unstoppable energy. She has been involved in restoring over 1,200 hectares (nearly 3,000 acres) in more than 100 peatlands across North America. The cost of restoration is between $3,500 and $4,000 per hectare—well within the range that peat harvesters can afford to pay.

It's a different story in the oil sands, where companies such as Syncrude and Suncor have brought in restoration experts like Jonathan Price, David Cooper, and Dale Vitt to help them reconstruct peatlands that have been peeled away for bitumen or degraded by roads, well pads, and seismic lines.

It wasn't civic duty that compelled them to so. Since its inception a half century ago, the oil sands industry has been getting bad publicity for the air and water it pollutes, for the caribou that have almost disappeared from peatlands it exploits, and for the fens, bogs and marshes it has destroyed or degraded. The scale of the degradation is staggering. Wetlands encompass about nearly 20,000 square miles, more than half of the oil sands region.[8] About two-thirds are bogs and fens. The rest is swamp, marsh, and open-water wetland. Once the mining for bitumen is completed, more than 110 square miles of peatlands will be lost.

It would have been a lot worse had the plan to use nuclear bombs to generate the intense heat needed to separate oil from bitumen gone ahead in the 1950s after getting initial approval from the US Atomic Energy Commission and Canada's Atomic Energy Department. That plan remained in play until 1976, when new and safer technologies for separating oil from bitumen came onstream. Safer technologies, of course, didn't prevent the environmental catastrophe that the oil sands would become in the three decades that followed.

Up until 2007, not a single acre of mined peatland had been certified as being "reclaimed" according to government standards. Just 1 percent was turned into forested uplands, shrublands, and swamps.

Most everyone thought that it was impossible to re-create peatlands in the oil sands, given how invasive the mining process is. First the landscape is drained or dewatered. Then the trees are chopped down before mineral overburden and layers of peat, two to three meters thick, are stripped away. The bitumen is then mined at depths of 150–300 feet before being taken to a refinery, where, until recently, oil was separated from the sand with the help of sodium hydroxide—a saline solution. What is left to reclaim are enormous pits that are filled in with that sodium-rich sand and leftover peat mix before seeding takes place. Instead of fens and bogs, upland forests and shrublands are created. While they looked good in places where bison were introduced to wild it up, they did not function as bogs and fens.

Scientists Suzanne Bayley and Rebecca Rooney compared 20 reclaimed peatland sites that were an average of 16 years old with 25 undisturbed sites in the oil sands area. They found that 70 percent of the reclaimed sites were in poor ecological health, with lower biodiversity, less-productive plants, and more land exposed to erosion and carbon loss.

Officials in the industry often countered bad news like this with promotional videos such as the one in which Greenpeace co-founder Patrick Moore walks through a restored peatland, extolling the beauty of the landscape. Never mind that Moore had long lost credibility in the conservation community. What stood out for those in the know who viewed the video was a bee buzzing around a bright yellow flower. Biologist Rooney recognized the flower as bird's-foot trefoil, an invasive species from Eurasia.

The news for the oil sands continued to get worse as ecologists such as Merritt Turetsky began to quantify the amount of carbon lost in mined peatlands as well as in the peat that was being stockpiled for reclamation. She concluded that most of the carbon in that peat would be lost to the atmosphere in fifty years.

Bringing in Dale Vitt in 2007 to oversee the reconstruction of a fen at Syncrude's Sandhill, an open-pit mine site, was a stroke of genius; not only was he Bayley's colleague and friend, and one of Turetsky's former teachers, he was also admired by David Schindler, a world-class limnologist and a relentless critic of the oil sands industry. Schindler viewed Vitt as one of the top wetland scientists in the world. Vitt, Schindler once told me, was one of the reasons he accepted an invitation to join the faculty at the University of Alberta.

Unlike Schindler, who peppered every interview with wry observations, cutting jabs at industry, and lively anecdotes such as the one where he nearly died in a bush plane when it crashed at the Experimental Lakes Area research site in northern Ontario, Vitt is much more low-key. He made no mention of being butted by a muskox in the Arctic or treed by an angry sea lion in New Zealand. Nor did he tell me that he did his graduate work with Howard Crum, the legendary "bryologists' bryologist" who wrote the definitive treatise on sphagnum, and who delighted in giving mosses names such as "The Extinguisher," "Knight's Plume," and "Goblin Gold"—a "singularly brilliant vegetable." (Goblin Gold is magically luminous for a plant that thrives in the darkest, dampest places.)

What Vitt liked to do most was get down on his hands and knees with a magnifying glass to look at mosses. His only critical observation came when he wondered why it was that there was only one peatland specialist working in the Alberta government when peatlands cover 40 percent of the forested area of the province.

Syncrude initially made much of the wetland that Vitt and his colleagues helped create at the Sandhill site, parading journalists in and declaring victory in the restoration effort. The media bought it. For his part, Vitt was more sanguine about the success of a ground-breaking experiment that began in a greenhouse to determine which peatland plants could tolerate high concentrations of sodium. *Carex aquatilis*, a water sedge, appeared to be the most promising.

Vitt looks back on it now and admits he was excited when the *Carex aquatilis* that dominated part of the fen started producing moss and carbon layers beneath it. But as cattails simultaneously took over the center of the wetland, and grass took over other parts, his expectations were tempered. Cattails do not form peat. Instead, they produce a wetland that will eventually become a cattail-dominated aquatic ecosystem that is unlikely to grow a lot of peat. Not bad, but not what they were aiming for.

"Everyone thought the engineers had designed the system so that the sodium would not be a problem, but it was," he said. "Mosses don't grow in high-sodium areas. They grow only in highly calcareous environments. It's a problem in attempting to restore this to the rich fens that once dominated this landscape. Those fens have high concentrations of calcium and magnesium and very low concentrations of sodium. If you add sodium to a fen, the mosses die. And remember, mosses form the majority of the peat column. If you don't have mosses, you don't have as much peat."

University of Waterloo peatlands expert Jonathan Price faced the same challenges when Suncor, another oil sands giant, brought him, David Cooper, and others in to construct a fen at a similar site. They have been making progress, such as identifying four mosses that can tolerate some level of salinity. Price allows that it is a challenge because of all the salty solutions that were used to extract bitumen in an already saline environment. Water is being discharged from the fen, but not fast enough to prevent the buildup of salt in the short term. Price is optimistic that half of that salt will be flushed out in about fifteen years.

"Engineering a landscape such as this requires a huge range of intellectual and academic disciplines," Price told me when I talked to him during the early stages of the restoration efforts. "Attempting to do it makes you appreciate just how complicated and precious this environment is and how difficult it is to re-create. But the fact is, resource extraction is inevitable. We're trying to make the best out of a difficult situation."

Rochefort has followed these developments in the oil sands with keen interest because they represent worst-case scenarios in peatland reconstruction. As optimistic as she is in believing progress will be made at the Suncor and Syncrude experimental sites, she suspects that it will be very expensive, possibly too expensive for an industry whose reclamation liabilities are growing at a time when profits are shrinking—even as expansion into virgin peatlands continues.

Nevertheless, even if the oil sands industry fails to restore peatlands back to fens and bogs, the research that is being conducted will not be a waste of time. The environmental catastrophe created by the oil sands extraction has brought together a veritable who's who of experts such as Vitt, Price, and David Cooper, hydrologist Kevin Devito, wetland ecologist Lee Foote, carbon-cycling specialist Elyn Humphreys, and Jan Ciborowski, an expert in aquatic ecotoxicology. They are working with soil scientists like Bing Si and hydrologist Sean Carey, as well as many engineers and land-reclamation experts. The last time I looked, more than seventy scientists and students had cycled through Price's research group, which is just one of many.

Talking to many of these scientists, I asked why it was that they were operating in such a low-key manner in such high-priced research. The problem, they all told me, was the oil sands, which is such a toxic subject that it's almost impossible to link a good news story such as the progress that is being made in growing peat to an environmental catastrophe that is taking place.

They are right. Figuring out how to grow and restore peatlands in the oil sands has implications for everything from commercial blueberry and peat moss harvesting to flood mitigation and climate change. But in continuing to mine the oil sands for bitumen, currently approved oil sands operations will release between 13 and 52 million tons of carbon. These changes will also reduce the former wetlands' ability to sequester carbon by as much as 8 million tons annually.[9]

"Like it or not, the reclamation debt is growing and it will continue to grow as oil sands companies move to expand production," says Bayley, who calculated those numbers with the help of her husband, Dave Schindler, and scientist Rebecca Rooney. "If this continues without a clear wetlands reclamation policy, we will have more than 65 percent less peatland and very little of the plant and animal life that existed there in the past."

Hans Joosten is optimistic that the world is beginning to recognize the importance of peat and the need to rewet those peatlands that have been disturbed or degraded, or are drying out because of climate change. But like Bayley, he fears that some of the research into growing peat is being used by industry as a social licence to continue doing what they are doing. "Peatlands are unique, complex, self-regulating [ecosystems] that are similar to those of an organism," he says. "Vegetation is just one part of their hierarchical structure. You can't measure the success of restoration by the number of plants that grow back. There's a lot more to peatlands than that. Their structure and function are dependent on so much more, such as water quality, water levels, temperature, humidity, and minerals. They are thousands of years in the making and they provide so much for biodiversity both within and outside their borders."

The conversation with Joosten got me thinking about the peatlands in Voyageurs National Park. Many of the wolves there have become experts in killing beavers, which is difficult to do because beavers are in the water most of the time and it is difficult to predict where they will emerge on shore. A four-year study led by scientist Tom Gable found that in an average year, the wolves, by killing beavers, have prevented the beavers from creating 88 ponds in a 772-square-mile area. This is important because beavers mine peat to build their dams, and the dams they build create pools of water that can slow or reverse the ability of a peatland to store carbon.

On the other hand, though, the fact that beavers continue to come back is important because their dams rewet peatlands that might otherwise dry out. It's the kind of water balance and water chemistry that is needed to construct a fen or restore one. And there are so many other pieces to the puzzle—precipitation and ground water, for example, and the right mixture of mosses and sedges—that add to the challenge when other factors such as roads, seismic lines, wildfire, drought, and flooding come into play.

The last peatland that Rochefort took me to was the 1,500-hectare (3,700-acres) Grande Plée Bleue, an 8,400-year-old, largely unexploited bog that is located twelve miles from Quebec City. The only signs of human intervention are the remnants of ATV trails and a

750-meter-long canal that was excavated in the 1950s in a failed attempt to turn this peatland into farmland. It was clear that this raised bog was special to Rochefort, not only because she was involved in the restoration of the drained areas, but also because she and others convinced the Quebec government to establish it as a protected area.

The restoration was only partially successful. But lessons learned from the failures shed new light on the limitations of sphagnum reestablishment in high cover shrubbery, as well as the spacing between dams and weirs necessary to restore hydrology.

As we sloshed along, it was hard not to be enchanted by the pink blooms of meter-high bog rhododendron, the exquisitely charming, albeit toxic sheepkill (*Kalmia angustifolia*), and the bell-shaped flowers of leatherleaf. Some species of dragonflies, I learned, are found nowhere else in the region. When we reached the highest point of the bog, there were larch *krummholz* (stunted, windblown trees) and many small ponds in a tundra-like setting. It reminded me of the floating islands of peat I had come across thousands of kilometers to the northwest in Macmillan Pass on the mountainous border of the Yukon and Northwest Territories.

The earth trembled beneath us as we walked on a thick carpet of semi-detached peat floating on cold, black water. One plant that stood out was a single pond lily that had not quite opened up. I had observed a lily like this before on the edge of the magnificent McClelland Lake fen that was being targeted for oil sands expansion. I like to think that it was the white waterlily or the Leiberg's waterlily that I saw there. But both are too rare for me to have been so lucky. Both are ranked S1 in Alberta, signifying that five or fewer occurrences are known in the province. One of them had been spotted at this fen.

I realized at some point that lilies and other rare plants in that part of Alberta—Herriot's sagewort, turned sedge, beaked sedge, larger Canadian St. John's wort, and pitcher plant—are in a precarious state, as many peatland species are. They are likely doomed if oil sands expansion proceeds, which the Alberta government seemed intent on when it allocated $30 million for the creation of the Canadian Energy Centre in 2020.

Peatland ecologist Line Rochefort in the Grande Plée Bleue, an 8,400-year-old, largely unexploited bog that is located 20 kilometers from Quebec City. Rochefort has restored more than a hundred bogs and fens. (Photo: Edward Struzik.)

There seems to be no giving up or acknowledging that there may not be a future for an industry that contributes so much to greenhouse-gas emissions. Those running the so-called Energy War Room have targeted industry critics—from environmental groups that get some of their funding from foreign sources to a Netflix children's show about a member of a Bigfoot family that questions whether there is such a thing as "clean oil." It was hard to fathom how an industry that was once confident about using nuclear bombs to produce the heat needed to separate oil from bitumen was so insecure that it would lash out at a children's show about a Bigfoot family.

Conclusion

Peat is an organic material that forms in waterlogged grounds, such as wetlands and bogs. . . . It can also be used to provide structural support and moisture for smooth, stable greens that leave golfers with no one to blame but themselves for errant putts.[1]

— Chief Justice of the US Supreme Court John Roberts in 2016, siding with three companies that challenged a US Army Corps of Engineers decision prohibiting them from harvesting 530 acres of peat moss from a bog in Minnesota for use in golf courses and landscaping

There is a sign above the door of a giant glass-enclosed bog chamber in a remote part of the Marcell Experimental Forest in Minnesota that explains why so many scientists from around the world have worked hard to get a piece of this wilderness. It is here in the southern boreal forest where they have the opportunity to speed up time to see how peatlands will respond to a rapidly warming climate.

SPRUCE—the Spruce and Peatlands Responses Under Changing Environments experiment—is a collaboration between the US Department of Energy's Oak Ridge National Laboratory and the US Forest Service. Each one of the ten chambers is 30 feet high, and 39 feet in diameter. All of them are designed to mimic what will happen to peatland ecosystems under various climate-change scenarios. They range from no change to a very realistic increase of +2.25°C, +4.50°C, +6.75°C, and finally a more frightening, albeit unlikely, +9°C spike (that is, a range of about +4°F to +16°F).

"Just about everything that's measurable is being measured," said US Forest Service scientist Randy Kolka as he walked me through one of several bogs and fens at Marcell before guiding me through the chambers. "Some of the best climate-change science in the world is being done right here in Minnesota. It's a one-of-a-kind experiment that will help us better understand how peatlands like this one will fare in a warming world and how much carbon they might store or release."

I made the drive to Marcell from the Oak Hammock Marsh just outside of Winnipeg, Manitoba, where the Canadian Air Force used to practice bombing because the marshes there were uninhabited and deemed to be wet and worthless. The military is still on the lookout for unexploded ordnance, and will be for some time in areas which the 100,000 annual visitors are not allowed to enter.

Just as it's easy to get lost sloshing through a marsh, bog, or fen, it's not hard to go astray on a long drive through northern Minnesota if you don't have a good map or a cell phone app telling you where to turn. Minnesota has 6 million acres (2.4 million hectares) of peatland, more than any other state outside of Alaska. There is no interstate highway running through them. The roads are marked, but sometimes in a manner that is confusing to visitors who are distracted, as I sometimes was, gazing at all the eagles, hawks, falcons, and osprey I saw along the way.

There is little to distinguish one mile of peatland from another in this part of the world unless you are in a plane and able to see the swirl of blue-green fens and black-tea bog streams slicing through islands of mounded peat. So many of them detach and float around with the wind that the state of Minnesota issues permits to companies and individuals to move the floating islands to places where they will do no harm.

Tamaracks, which grow in peatland forests and on some of these floating islands, need 300 years to get seven feet high. The trees have lived that long only because loggers who combed through this region saw no commercial value in felling them. As worked over as they have been, the peatlands are still inhabited by wolves in places like Voyageurs National Park, where some have learned to prey on beavers. Marcell has more back bears per acre than any other place in the United States except maybe for the Great Dismal Swamp and the Pocosin Lakes National Wildlife Refuge.

Northern Minnesota is much like Manitoba in that most every-

US Forest Service scientist Randy Kolka in one of ten chambers in the Marcell Experimental Forest of Minnesota that are designed to mimic what will happen to peatland ecosystems under various climate-change scenarios. (Photo: Edward Struzik.)

one seems to be as nice as the stereotypical image of people living in that state. A mean-looking, leather-faced pickup driver lit up with a smile when I reluctantly pulled over to ask for directions while he was changing a tire. Two young women with rainbow-colored Mohawk haircuts were only too happy to help when I could not find my way out of Grand Rapids to get on the road to Marcell. The graduate student who was overnighting at the US Forest Service Station at Marcell talked my ear off telling me all about the black bears he was tracking.

Kolka was not out of character when he arrived the next morning and greeted me heartily. He is as stout as the wrestler he once was in school. The first photo on the PowerPoint he presented as part of my orientation was one of him proudly holding an enormous muskellunge, a fish that looks like a big pike, but is so difficult to catch that it is known in these parts as "the fish of a thousand casts." *Water wolf* is another name for it.

Since 2016, when both heat and carbon dioxide first started to be pumped into these chambers, scientists have been monitoring changes in soil alkalinity and specific conductivity, as well as the total phosphorus, nitrate, and phosphate in the soil. They've been looking at the chemistry of peat, the genetics of bacteria, and the amount of carbon dioxide and methane that is being stored or emitted.

As Kolka and I went from one chamber to another, starting from no change in temperature to the one in which it is nine degrees Celsius warmer than normal, it was easy to visualize how peat-loving plants respond to extreme increases in temperature and CO_2. "The tamarack and the spruce don't look good," said Kolka. "But the shrub layer is doing really well. It's so dense that you can hardly see the sphagnum. Those mosses are not going to last. That's the bad news for peat."

Marcell is an ideal place to conduct a study like this because it's about as far south as boreal peatlands extend in North America, and it has been in transition from coniferous forest to hardwoods because of climate change, wildfire, and past forestry practices. The peat continues to store fifteen times more carbon than the recut forests of the East, but this ecosystem is on the verge of flipping, at which point it will send more greenhouse gases into the atmosphere than the amount that peat typically stores.

After just three years of tracking changes in plant growth, water and peat levels, microbial activity, fine-root development, and other things that control the movement of carbon into and out of the glass-enclosed bog chambers, Kolka and Paul Hanson, an ecosystem scientist at Oak Ridge, have found that the bogs in the warmed chambers are quickly making the transition from being carbon accumulators to carbon emitters. Even those that were warmed modestly lost carbon five to twenty times faster than historical rates of storage.

The only thing quite like it on this scale in North America is the Experimental Lakes Area project in northern Ontario, which was established in 1966 to deal initially with algal blooms on the Great Lakes before expanding to study the effects of acid rain, hormones, mercury, pharmaceuticals, and plastics on freshwater ecosystems.

The idea of damming, polluting, and acidifying dozens of lakes back then was considered to be radical, and to some unacceptable. But in fertilizing some pristine lakes with phosphorus and nitrogen, and adding minute amounts of sulfuric acid to others, ways were found to deal with eutrophication and acid rain, which were killing fish and other aquatic species. The data, and especially the photos, were so compelling that conservative leaders like US president Ronald Reagan and Canadian prime minister Brian Mulroney jointly agreed to ban phosphates from detergents and order industries to limit sulfur dioxide emissions in the 1980s.

Given its tiny footprint, SPRUCE is not nearly as obtrusive as the Experimental Lakes Project. But the results of these experiments will be just as useful in informing decision-makers on the best ways of managing peatlands in the future.

There are lot of ideas about how to do this in a world in which temperatures and carbon dioxide levels are rising rapidly. According to data compiled by Hans Joosten and others at the Greifswald Mire Centre in Germany, peatland emissions have been rising in fifty countries since 1990. Excluding emissions from wildfires, Indonesia and the European Union (which included Great Britain at the time of their report in 2018) led the pack, with Russia, China, and Malaysia vying for third place and the United States not far behind.

Geoengineering ideas tend to get the most attention because of the high cost and sci-fi nature of managing solar radiation by sending aerosols to the stratosphere to dim the rays of the sun, or by kickstarting the growth of sea ice by spreading sunlight-reflecting silica beads on the Arctic Ocean. Harvard scientist David Keith, one of the key players in SCoPEx, a scientific experiment to advance our understanding of stratospheric aerosols, has made the case for solar management. He and others believe that "cutting emissions to zero tomorrow does not deal with the climate risk."

"There is evidence that a combination of emissions cuts and solar geoengineering might be significantly safer than emissions cuts alone," he told a group of scientists, politicians, policy makers, and indigenous leaders in a series of online presentations in 2021 organized by scientist Duane Froese and the privately funded organization Permafrost Carbon Action.

I found it a compelling argument as I listened in on the presentations, but they did not take into account what this will do to plants that rely on direct sunlight or to Arctic phytoplankton that rely on solar radiation being diffused through sea ice. Phytoplankton are food for krill, which are food for Arctic cod, which provide the fat and protein that sustains seabirds, seals, belugas, and narwhals.

We have seen what happens naturally when volcanoes spew aerosols into the stratosphere and dim sunlight. The eruption of Mount Pinatubo in 1991 resulted in a short-term cooling and near complete reproductive failure for waders and waterfowl throughout the Arctic the following year.[2]

Pam Pearce, director of the International Cryosphere Climate Initiative, articulated some of these concerns that day at the same session. "We should not mislead ourselves that we can solve these problems, or take huge risks of creating new ones," she said, stressing what she called "the importance of engaging local communities, and especially indigenous communities meaningfully in these kinds of discussions."

The relevance of her suggestion was underscored just a few days later when the Swedish Space Corporation, under pressure from indigenous people and environmental organizations, called off a landmark first test of the technology that Keith and the SCoPEx project were about to undertake in northern Sweden. Åsa Larsson Blind, vice president of the Saami Council, said in a statement that such technological fixes were "completely against what we need to do now—transform to zero-carbon societies in harmony with nature."

There are other geoengineering experiments that have been put into play, with mixed results. Technology that strips out carbon dioxide from the exhaust gases of industrial processes before it is liquified and pumped into the ground has proven to be expensive as well as disappointing, especially when the captured carbon is used to pump more oil out of the ground.

Another idea that has been put forth is bioengineering landscapes to slow the thawing of permafrost and the shrubification of tundra wetlands. Sergey and Nikita Zimov's Pleistocene Park is an intriguing example, if only because it is so off-the-wall and receives so much adoring media attention. The Zimovs are a father-and-son science team who created the 50-square-mile nature reserve in eastern Siberia in 1996 with the long-term goal of returning much of the landscapes of northeastern Siberia, Alaska, and the Yukon to the grasslands that existed 12,000 years—the end of the Pleistocene Epoch. The idea, financially supported, in part, by crowdfunding—is to populate them with caribou or reindeer, muskoxen, bison, and even mammoths, should a way be found to bring them back to life. The theory is that the animals will trample the soft snow into ice and lock up carbon that is being released by rapidly thawing permafrost for longer than it is locked up now with just snow covering it for a good part of the year.

The Zimovs' contention that animals such as reindeer will slow or stop the shrubification of the Arctic isn't without merit. Poring over three decades of satellite imagery of vegetation in the reindeer herding area in the Yamal Peninsula in northwest Siberia, scientist Megha Verma and her colleagues found no evidence of shrubs expanding in the study site.[3]

Pleistocene Park is flawed in a number of ways, however, because it assumes that indigenous people, not climate change, were mostly responsible for killing off the now-extinct grass- and sedge-feeding animals. That would have been an impossible feat for a small, scattered population that had no guns or motorized vehicles to hunt the animals down.

The concept also neglects to take into account how things might backfire, as they did in the wetlands of the Kaibab Plateau north of the Grand Canyon when bison were introduced as a cattle–bison breeding experiment. There is some peat there, though not a lot of it. And there are floating islands, too. Vegetation in the Kaibab Plateau wetlands, which includes dozens of rare plants, was so badly degraded in the years following the introduction of bison that there was close to a 50 percent increase in bare soil.[4]

It's also difficult to reintroduce animals, except possibly bison, to a polar world, even from one part of the Arctic to the other. The Canadian government learned this lesson when they tried to introduce reindeer from Alaska to the Tuk Peninsula in the 1930s. It would be even more difficult now, given the fact that biting flies, snow cover, icing events, disease, and predators are keeping ungulates in check even under the best of conditions. Many of those once-pristine places are fragmented by roads, mines, and oil and gas development that make it difficult for shy, climate-challenged animals such as caribou to move from one wilderness to another. These are some of the reasons (along with overhunting) that most of the world's caribou populations are in freefall. The Bathurst herd in Canada's Central Arctic dropped from roughly a half-million animals in the 1980s to a low of just over 8,000 when they were counted in 2018.

Big ideas like Marcell and a research tunnel that was bored into the permafrost in Alaska back in the 1960s are necessary because they allow scientists to both speed up time and turn the clock back to the days when steppe bison and woolly mammoths roamed the tundra. The original purpose of the tunnel at the Fairbanks Permafrost Experiment Station was to test the design and development of new construction and road-building techniques in cold climates. But research has opened up to other disciplines, leading to serendipitous discoveries such as microscopic ice-nucleating particles in permafrost that could affect Arctic clouds if they become airborne.

The US Army Corps of Engineers' $50-billion plan to divert freshwater and sediment from the Mississippi River into the depleted wetlands of Barataria Basin in Louisiana is a gamble, given how quickly sea levels are rising. It is designed to reverse the subsidence that occurred when levees were built to control water flows. The aim is to accumulate peat and sediment on the marsh bottoms so that 17,300 acres of new land will rise up in thirty years. The fact that it is likely to have a negative impact on dolphins, shrimp, and other aquatic species has been recognized. As Chip Kline, chairman of the Louisiana Coastal Protection and Restoration Authority, explained: "This is what climate adaptation looks like at scale."[5] It could be a win-win outcome if it results in a healthy coastal habitat. As scientists with the National Oceanic and Atmospheric Administration are beginning to discover, coastal salt marshes, mangrove forests, and sea grass meadows are remarkably efficient—much more so than mature tropical forests—at capturing and storing large amounts of carbon.

The risk of investing so much money in big ideas is that it could draw attention away from simpler, more cost-effective solutions. "Low-hanging fruit" is how wetland ecologist David Locky described it back in 2011 when he suggested that Alberta and other Canadian provinces were at a crossroads, with the degradation of peatlands, especially those in the oil sands, rising at a rate that was rapidly overtaking the ability to economically restore them back to life and functionality. It was a prescient call, because two of the Alberta's worst natural disasters—the $6-billion Calgary flood of 2013 and the $8.8-billion Horse River fire of 2016—were more damaging than they should have been because of wetland and peatland degradation.

There is still a lot of low-hanging fruit.

Scientist Thomas Dreschel found a promising way of regrowing trees on the rapidly drowning, floating tree islands of the Everglades, which serve as refuges for birds, plants, and land animals and as filters for removing nutrients that flow in from nearby farms. All it took was a small amount of sphagnum moss from Canada to serve as a growing medium for the saplings. Nothing else had worked up until then.

Strategically managing wildfire in the boreal forest and permafrost areas in association with indigenous communities in order to protect rich carbon reserves is also an idea worthy of consideration, as scientist Merritt Turetsky and others have suggested. "Peatland carbon sequestration requires fire—not frequent fire, not severe fire, like we're starting to see," Turetsky said when she addressed the Permafrost Carbon Action Group in 2021. "But low levels of burning. It opens up the landscape. It keeps sphagnum moss in place, and not being taken over by other kinds of dry or non-carbon-accumulating ground cover like lichens."[6]

Reintroducing beavers to rewet badly degraded peatlands, as was done near Blairgowrie in eastern Scotland in 2002, is an ingeniously simple idea that could augment strategic wildfire management in the boreal world. The Scottish fen was drained in the 1800s to create pasture for livestock. This was long after beavers had been hunted and trapped to extinction. It's not the only time this has been done. Beavers were smuggled into the River Tay for the same purpose, and with so much hand-wringing and finger-pointing that an official with the Scottish National Heritage warned that "the presence of beavers on the Tay undermines our credibility as a country in handling these things properly, legally, and democratically."[7] The rodents got to stay and redeem their reputation. After just ten years, the beavers near Blairgowrie restored the ecosystem better than any form of human intervention.[8]

The US Department of Agriculture's Wildlife Services program kills more than 22,000 beavers each year. Redistributing beavers to even out moisture may be a better way of dealing with these so-called problem beavers and restoring and maintaining peatlands at the same time. The US Department of the Interior did this in Idaho during the Second World War. Initially, the animals were transported into remote areas on the backs of mules. But when it became evident that mules and beavers didn't get along, the beavers were parachuted in, led by a beaver named Geronimo, who made several test flights. It worked. Oregon did the same when it removed hundreds of so-called bad beavers from the lowland farm districts to the high country, where they would be of "inestimable value in water conservation, flood, and erosion control."[9]

In 2014, the US Forest Service began reintroducing beavers to the Scott River basin in California, which was called Beaver Valley before the animals were trapped out with the notable help of one man who killed 1,800 beavers in a single month in 1836.[10] The case study was so successful that other districts in the state are following suit.

There is the danger of beavers running amok, as they did when Canadian beavers were relocated to the peatlands of Tierra del Fuego in 1946. The newcomers had no business being in an environment where there were no predators to keep them in check and in forests that had not adapted to periodic flooding. In 2020, scientists Alejandro Huertas Herrera and his colleagues reported more than 200,000 beaver dams flooding 28,000 square miles of peatlands. Not only is the flooding killing trees, but it is also releasing huge amounts of methane that was safely secured in the undisturbed peat.[11]

There is a Chinese proverb that suggests that "the best time to grow a tree is twenty years ago. The second best time is now."

Great efforts to plant trees are being made throughout many parts of the world. In December 2020, the Canadian government announced that is investing more than $3 billion to plant two billion trees over the next ten years. It's difficult to take issue with that, because trees cleanse the air, filter water, and keep nutrients in the soil. Trees also absorb carbon through photosynthesis. But dollar for dollar, a climate-mitigation investor would get bigger returns by growing or accumulating peat, rewetting peatlands that have been degraded, and conserving the healthy peatlands that still exist. There may even be merit in removing trees such as black spruce in peatlands to help restore or maintain the moisture of the peat. Restoring peatlands is 3.4 times less nitrogen-costly than planting trees, because additional nitrogen is needed to build up a similar carbon pool in organic matter of mineral soils.[12]

Peatlands, which represent no more than 4 percent of the earth's terrestrial surface, store twice as much carbon as all of the world's forests. They not only filter water that yaks, cattle, sheep, and fertilized farmlands pollute, but they also hold back that water in times of flood and store it during extended drought.

Many countries have been successfully rewetting degraded peatlands for a variety of reasons. For Russia and Indonesia, and for the US Fish and Wildlife Service that manages wildlife refuges in the Pocosins and the Great Dismal Swamp, it is a way of curtailing peat fires that have become a serious public health problem. For Holland, it's mitigating the flooding and subsidence that is destabilizing homes and buildings. For Scotland, it's undoing the ecological damage that was done in the 1960s and 1970s when generous tax incentives encouraged timber companies to plant coniferous trees on the country's wondrous blanket bogs. And for China, the rewetting of the Zoigê Wetland is a water security issue for millions of people living downstream and for yak herders struggling to make a living.

Carbon farms, such as the one Duke University launched next to the Pocosin Lakes National Wildlife Refuge are also promising. Carbon farming involves rewetting dried-out or degraded peat to the point at which it is no longer releasing carbon. The owner of the 10,000-acre carbon farm near the Pocosin wants to remain anonymous. If this experiment succeeds in storing carbon, he and Duke University could get carbon credits that could be sold to other interests. The problem with carbon credits is that they allow energy companies, airlines, and the shipping industry that buy them to continue to rely heavily on fossil fuels.

None of this is going to be enough if countries like Ireland, Finland,

and Estonia continue to burn peat for energy. Peat extraction in Finland is generating almost 24 million tons of carbon dioxide per year, more than twice the emissions of cars, trucks, boats, and locomotives do in that country.

Ireland has vowed to phase out peat as a primary source of energy by 2030. But poor countries like Rwanda and Burundi are moving in the other direction because banks and international funding agencies aren't incentivizing them enough to make the transition to solar and wind. In 2020, a Congolese company was making a case for drilling for oil and gas in the Cuvette Centrale, that peatland the size of England that was not discovered until 2017.[13]

What's also needed is a serious environmental accounting, a robust concept I was introduced to a number of years ago when I was invited to speak about the true cost of wildfire to the North American Congress of Social and Environmental Accounting. On one simple level, environmental accounting measures the value of a natural resource such as peat when it is extracted to condition soil or to provide fuel to power energy plants. On another, more complex level, it assesses the real costs of exploitation and restoration and the true value of a fully functioning ecosystem like a peatland that stores carbon, filters water, mitigates flooding, slows wildfire, and provides refuges for endangered species. The tricky and much neglected part of this equation is how to measure the functionality of a restored or reconstructed ecosystem.

In the past, forestry, mining, and energy companies as well as real estate developers made concessions to conservation interests by offering substitution rather than restoration. Syncrude, an oil sands giant, put bison on a fen it had mined for bitumen, declaring the restoration a resounding success when the herd increased from 30 to 300 animals. The carbon sequestration and species diversity associated with the peatland fen that was there before was lost, but no one cried foul because bison grazing on grasslands looked pretty good. Coal-mining companies do the same when they plant alfalfa, clover, or Kentucky bluegrass on mined-out mountain sites. No one complains, because the succulent vegetation attracts bighorn sheep, deer, and other animals.

Canadian scientist Scott Nielsen and Federico Riva, a postdoctoral fellow in Nielsen's lab, underscored the complexity of assessing ecosystem functionality in a study that showed that even where resource development in the boreal peatland forests of northern Alberta did not result in the loss of a significant amount of trees, ecosystem functionality varied broadly and often fared poorly for species habitat, biodiversity, and the increased risk of wildfire ignition.[14]

Peatland restoration is necessary because, left to nature, degraded bogs and fens have a tough time coming back, as Cassidy van Rensen, another one of Scott Nielsen's students, found out when she looked at how forested fens regenerated naturally along oil and gas seismic lines that had been cut through the fens in the oil sands region. Even after a half century, many of the fens failed to come back.[15] What they did do, according to scientist Maria Strack, was blow off a lot of methane, a potent greenhouse gas.[16] It's an important consideration because wetlands cover more than half of Alberta's oil sands, with forested peatlands accounting for over 90 percent of these wetlands.

Too often, the measure of the success of restoration is based on the trees, smaller plants, and large charismatic animals on top of the ground rather than on what is below, very close to the surface, or buzzing around in the air.

Restoration projects tend not to consider the return of invertebrates such as soil-conditioning ants; pollinating bees, moths, and butterflies; spiders that regulate the density of other invertebrate herbivores; and beetles that foster and support the decomposition that is mostly done by fungi and bacteria. As they tunnel into decaying wood, beetles and their larvae set up "entry courts" for the fungi to get in, according to entomologist John Spence, my go-to expert on such subjects. Some beetles, such as bark and ambrosia beetles, have symbiotic relationships with fungi, purposely bringing spores into the wood in special pouches called *mycangia*. Ambrosia beetles feed on the fungi that in turn grow in their galleries by decomposing wood. Bark beetles improve fungi survival by cutting tree ducts that bring resin to suppress the growth of the invaders.

Very few projects, if any, look at what happens to the 601 species of fungi, the dominant microbes in many acidic peatland ecosystems. It's an important consideration, because fungi also have a powerful relationship with orchids and other plants that grow in peatlands. As facilitators of decomposition, they play an important role in carbon mobilization and storage. And by linking the roots of different plants, the mycelium of fungi allow the plants to share nutrients or shield themselves from toxic chemicals that invasive plants might bring.

"We are just starting to understand what happens to things like spiders, beetles, and ants once a restoration treatment is applied," Jaime Pinzon, a research scientist with the Canadian Forest told me. Pinzon is attempting to make a more holistic assessment of functionality in restored peatlands in northern Alberta. "But given the scarcity of baseline information, it is very challenging to judge on the short term whether assemblages in restored areas will eventually converge toward those typical of undisturbed conditions."

Rarely is any of this accounted for in megaprojects such as the $9-billion Site C hydroelectric project in British Columbia and the $13-billion Muskrat Falls development in Labrador that will degrade and flood peatland ecosystems. Nalcor, the government-owned energy company behind Muskrat Falls, stated that there were just too many bogs and fens to account for when it submitted its environmental impact statement for regulatory review. Benga, the coal-mining company that plans to lop off mountaintops in the Southern Rockies to ship coal to China and India, concedes that it will destroy mountain fens without acknowledging how it will affect groundwater.

It's remarkable that the United States provides billions of dollars in direct subsidies annually to the fossil-fuel industry ($24 billion in 2019),[17] but leaves it up to a handful of scientists to deal with rewetting peatlands such as those in the Great Dismal Swamp and the Pocosin, and that the US government can't find a million dollars for the Plant Extinction Prevention Program in Hawaii, which is attempting to save the 220 species of plants that have fewer than 50 specimens remaining in places such as the Alaka'i Swamp.

The story of peatlands is sobering, given how much has been lost and degraded. But the future of peatlands does not need to be as melancholy and hopeless as the melting of Arctic and Antarctic sea ice and the meltdown of the Greenland ice cap. Little can be done to grow ice or save what's left of it without massive investments that will likely fail or even backfire. But much can be done to grow (that is, accumulate) peat, rewet degraded peatlands, preserve those that remain pristine, and reintroduce critically endangered species such as the red wolf and the red-cockaded woodpecker to them.

There is, however, more to it than reintroducing endangered species, capturing and storing carbon, filtering water, mitigating flooding, slowing wildfires, and providing refuge for so many plants, insects, and larger animals. Peatlands are also a cultural resource, a backdrop to poems, stories, and songs, the source of the sweet scent of peaty smoke-infused Scotch whisky, a record of the past, a reminder of the horrors of slavery and concentration camps, and the inspiration for mythical stories that remind us that wilderness can be warm and embracing just as much as it can be sublime, dangerous, and spooky.

I returned to my field notes recently to go over that sequence of events that led me to the cool, clear peaty stream I found when I needed it most during a 66-day kayak trip down the Nahanni, Liard, and Mackenzie Rivers in Canada's western Arctic.

Reading the words, I could visualize once again how the water sparkled in the sunlight and quenched my thirst, and how it smelled sweet like the whisky I had brought along. Was the flower I saw really a rare dragon's mouth orchid, or was it a more common calypso, which is just as beguiling and certainly not a disappointment?

What I had forgotten was the story of the giant thunderbird that was told to me when I stopped in at the Sahtú village of Fort Good Hope, where 850 people live in a place that has no permanent road linking it to the outside world. The thunderbird was a water monster that hunted down and killed both animals and people who passed through the wetlands until an elder with powerful medicine killed it. Unconvinced that the thunderbird is really gone, many people continue to conduct protective rituals when they boat or snowmobile through this area.

After reading that entry, I was curious to find out more about the peaty wetland that presented itself to me before that grizzly bear forced me to pack up and find another place to camp downstream. I was surprised to discover that Stephen Kakfwi, a Dene leader I knew when I lived in Canada's Northwest Territories, was working on behalf of the

Sahtú people of the region to set the wetland aside as an indigenous protected area. The last time I saw Steve, he was hunting caribou along the Keele River south of the Ramparts while my wife and I were paddling there.

When the deal is done, Ts'ude niline Tu'eyeta, known in English as the Ramparts River and Wetland, will ensure that 10,000 square kilometers of bogs, fens, and other peatlands and wetlands will still be there in a hundred years and more. It shuns the idea that its oil and gas resources, estimated to be vast, should be exploited. It shares what Gwich'in in Old Crow feel about the Old Crow Flats, and how the Mushkegowuk Cree see the Hudson Bay Lowlands; this spongy, obstructive, ambiguous, buggy wilderness that is as much water as it is land is essential to preserving their traditional food supplies, their identity, and the history and legends that have shaped their culture. Not a dollar will be made from it.

Notes

Preface

1. Charles Kingsley, *The Water-Babies, A Fairy Tale for a Land-Baby* (London: MacMillan, 1922).

2. Parks Canada, "Aulavik National Park—3.4 Ecological Integrity Statement, 3.4 Vision Statement, modified March 31, 2017, https://www.pc.gc.ca/en/pn-np/nt/aulavik/info/plan/plan2/sec3/page3.

3. B. J. Nicholson et al., "Peatland Distribution along a North-South Transect in the Mackenzie River Basin in Relation to Climatic and Environmental Gradients," *Vegetation* 126, no. 2 (1996): 119–33, www.jstor.org/stable/20048744, accessed February 1, 2021.

Introduction

1. Donald S. Murray, "Sphagnum Moss," from *The Guga Stone: Lies, Legends and Lunacies from St Kilda* (Edinburgh: Luath, 2013).

2. "Red River Settlement," *The Globe* (Toronto), February 16, 1869; ProQuest Historical Newspapers: *The Globe and Mail*, 3.

3. Smith Henderson, *Fourth of July Creek* (New York: HarperCollins, 2014), 206.

4. Kjeld Holmen and George W. Scotter, "Sphagnum Species of the Thelon River and Kaminuriak Lake Regions, Northwest Territories," *The Bryologist* 70, no. 4 (1967): 432–37, www.jstor.org/stable/3240785, accessed January 6, 2021.

5. William A. Niering, "Endangered, Threatened, and Rare Wetland Plants and Animals of the Continental United States," in *The Ecology and Management of Wetlands*, vol. 1, *Ecology of Wetlands*, ed. D. D. Hook (Portland, OR: Timber Press, 1988), 227.

6. International Union for the Conservation of Nature, "Issues Brief: Peatlands and Climate Change," https://www.iucn.org/resources/issues-briefs/peatlands-and-climate-change.

7. Halldór Laxness, *Independent People*, tr. J. A. Thompson (New York: Vintage International, 1997), 188.

8. Ibid., 192.

9. Ibid., 193.

10. Charles Lever, *Lord Kilgobbin* (London: George Routledge and Sons, 1872), 1.

11. Lynn Faust, *Fireflies, Glow-worms, and Lightning Bugs* (Athens, GA: University of Georgia Press, 2017).

12. Muiris O'Sullivan and Liam Downey, "Turf-Harvesting," *Archaeology Ireland* 30, no. 1 (2016): 30–33, www.jstor.org/stable/43745953.

13. William Howard Hearst (premier of Ontario), quoted by George Castle, "Will Explore Peat Lands," *The Globe* (Toronto), January 25, 1918.

14. The words were written by Johann Esser, a peat bog miner, and Wolfgang Langhoff, an actor. The music was composed by Rudi Goguel, a German resistance fighter.

15. William King, "Of the Bogs and Loughs of Ireland," Presentation to the Dublin Society, 1685.

16. Ralph Richardson, *On Peat as a Substitute for Coal* (Edinburgh: Adam and Charles Black, 1873), 10.

17. John Bunyan, *The Pilgrim's Progress*, edited with an introduction by Roger Sharrock (Harmondsworth, UK: Penguin Books, 1965), 46.

18. H. S. Joosten, M.-L. Taipo-Biström, and S. Tol, *Peatlands—Guidance for Climate Change Mitigation through Conservation, Rehabilitation, and Sustainable Use*, 2nd ed., *Mitigation of Climate Change in Agriculture* series (Rome: Food and Agriculture Organization of the United Nations, 2012).

19. Outi Ratamäki et al., "Framing the Peat: The Political Ecology of Finnish Mire Policies and Law," *Mires and Peat* 24. (2019): 1–12, doi:10.19189/MaP.2018.OMB.370.

20. Theophile Mugerwa, Digne Edmond Rwabuhungu, Olugbenga A. Ehinola, Janviere Uwanyirigira, and Darius Muyizere, "Rwanda Peat Deposits: An Alternative to Energy Sources," *Energy Reports* 5 (2019): 1151–55, doi.org/10.1016/j.egyr.2019.08.008.

21. Extrapolated from sources that point out that a 13,500-horse French army cavalry needed 22,000 tons of peat in a year. See: H. D. Joosten et al., "Wise Use of Mires and Peatlands," International Mire Conservation Group and International Peat Society, Helsinki, Finland, 2002, 56.

22. Timothy Winegard, *The Mosquito: A Human History of Our Deadliest Predator* (New York: Penguin Random House, 2019).

23. Nalcor Energy, "Habitat Types and Non-Habitat Areas within the Transmission Corridor for Central and Southeastern Labrador," 2007, https://muskratfalls.nalcorenergy.com/wp-content/uploads/2013/03/Chapter-10-Part-2A.pdf.

24. Peat can be as deep as ten feet in a few places in Georgian Bay.

25. Catherine La Farge, "Regeneration of Little Ice Age Bryophytes Emerging from a Polar Glacier with Implications of Totipotency in Extreme Environments, *PNAS* 110, no. 24 (June 11, 2013): 9839–44, doi.org/10.1073/pnas.1304199110.

26. Monika Ruwaimana et al., "The Oldest Extant Tropical Peatland in the World: A Major Carbon Reservoir for at Least 47,000 years," *Environmental Research Letters* 15 (2020): 114027.

27. G. J. Symons, "The Floating Island in Derwentwater," *Nature* 40 (1889): 290–91, doi.org/10.1038/040290a0.

28. Michael Oliver and Kenneth McKaye, "Floating Islands: A Means of Fish Dispersal in Lake Malawi, Africa," *Copeia* (1982): 748, doi:10.2307/1444082.

29. Sidney Powers, "Floating Islands," *Popular Science Monthly* 79 (September 1911).

30. US Department of Agriculture, " Assessing Proper Functioning Conditions for Fen Areas in the Sierra Nevada and Southern Cascade Ranges in California—A User Guide," April 2009, https://www.fs.usda.gov/Internet/FSE_DOCUMENTS/stelprdb5385279.pdf.

31. J. M. Glime, "Birds and Bryophytic Food Sources," Chap. 16-2 in: *Bryophyte Ecology*, vol. 2, *Bryological Interaction* (eBook) (Houghton, MI: Michigan Technological University and the International Association of Bryologists, 2017).

32. J. O. Rieley, "Biodiversity of Tropical Peatland In Southeast Asia," Abstract No. A-213, 15th International Peatland Congress, 2016, https://peatlands.org/assets/uploads/2019/06/ipc16p707-711a213rieley.pdf.

33. Cited from: M. E. Harrison et al., "Tropical Forest and Peatland Conservation in Indonesia: Challenges and Directions," *People and Nature* 2, no. 1 (2019): 4–28, doi .org/10.1002/pan3.10060.

34. G. Dargie et al., "Age, Extent, and Carbon Storage of the Central Congo Basin Peatland Complex," *Nature* 542 (2017): 86–90, doi.org/10.1038/nature21048.

35. L. I. Harris et al., "Lichens: A Limit to Peat Growth?" *Journal of Ecology* 106, no. 6 (2018): 2301–19, doi.org/10.1111/1365-2745.12975.

Chapter 1

1. George Percy, "Observations Gathered Out of a Discourse of the Plantation of the Southerne Colonie in Virginia," in *Hakluytus Posthumus, or Purchas His Pilgrimes*, compiled by Samuel Purchas (London: H. Fetherston, 1625), 1686–89.

2. College of William And Mary, "Extreme Droughts Played Major Role in Tragedies at Jamestown, 'Lost Colony,'" *ScienceDaily*, April 28, 1998, www.sciencedaily .com/releases/1998/04/980428075409.htm.

3. Esri, "Typhoid Fever," n.d., https://www.arcgis.com/apps/MapJournal/index.ht ml?appid=e8578e5bcf8244eb9daf771f9f386dd0.

4. William Byrd, "Description of the Dismal Swamp and a Proposal to Drain the Swamp" (1728), Electronic version, Library of Congress, https://www.loc.gov/resource /lhbcb.22884/?st=gallery.

5. John Smith, *The Generall Historie of Virginia, New-England, and the Summer Isles* (Bedford, MA: Applewood Books, 2006), 76.

6. William Byrd, *The Secret Diary of William Byrd of Westover, 1709–1712*, ed. Louis B. Wright and Marion Tinling (Richmond, VA: Dietz Press, 1941).

7. Byrd, *Description of the Dismal Swamp*.

8. "Resolutions of the Dismal Swamp Company, 1 May 1785," *George Washington Papers at the Library of Congress, 1741–1799, Series 9, Addenda to the George Washington Papers, 1763–1797*.

9. Moses Grandy, *Narrative of the Life of Moses Grandy, Formerly a Slave in the United States of America* (Boston: Oliver Johnson, 1844), 22.

10. James Redpath, *The Roving Editor: Or, Talks with Slaves in the Southern States* (New York: A. B. Burdick, 1859).

11. Marcus P. Nevius, *City of Refuge: Slavery and Petit Marronage in the Great Dismal Swamp, 1763–1856* (Athens, GA: University of Georgia Press, 2020), 21.

12. Dan Sayers, *A Desolate Place for a Defiant People: The Archaeology of Maroons, Indigenous Americans, and Enslaved Laborers in the Great Dismal Swamp* (Gainesville, FL: University Press of Florida, 2014).

13. John Ferdinand Smyth, *A Tour in the United States of America*, vol. 11 (London: G. Robinson, 1784), 102.

14. Dave Hunter Strother, "The Dismal Swamp," *Harper's New Monthly Magazine*, vol. 13, 1856, 453.

15. Frederick Douglas, "The Dismal Swamp," *The North Star*, March 11, 1859.

16. Redpath, *The Roving Editor*, 293.

17. Ibid., 291.

18. William Robinson, *From Log Cabin to Pulpit, or, Fifteen Years in Slavery* (Eau Claire, WI: James H. Tifft, 1913; electronic edition, University of North Carolina at Chapel Hill, 1997).

19. Ibid.

20. Frederic Law Olmsted, *A Journey in the Seaboard Slave States* (New York, Mason Brothers, 1861), 152.

21. Frederick Law Olmsted, "The South: Letters on the Productions, Industry, and Resources of the Southern States," *New York Daily Times*, April 23, 1853, https://proxy.queensu.ca/login?qurl=https%3A%2F%2Fwww.proquest.com%2Fdocview%2F95838950%3Faccountid%3D6180.

22. Moses Grandy, "Life of Moses Grandy," in *North Carolina Slave Narratives*, ed. D. A. Davis, T. Evans, I. F. Finseth, and A.N. Williams (Chapel Hill, NC: University of North Carolina Press, 153–83).

23. Olmsted, "The South."

24. Ibid.

25. Ibid.

26. Ibid.

27. Charles Frederick Stansbury, *The Lake of the Dismal* (New York: A. & C. Boni, 1925).

28. Robert Bransfield, "Lyme Disease and Cognitive Impairments" (webpage, n.d.), http://www.mentalhealthandillness.com/Articles/LymeDiseaseAndCognitiveImpairments.htm.

29. H. T. Crittenden, "The Dismal Swamp Railroad Company," *The Railway and Locomotive Historical Society Bulletin* 64 (1944): 61–68, www.jstor.org/stable/43517392, accessed September 14, 2020.

30. Ibid.

Chapter 2

1. Pehr Kalm, *Travels in North America*, 1750; quoted in: "New York Woodlands," *New York Nature* (website), 2020, https://www.newyorknature.us/woodlands/.

2. The image comes from a wonderful reconstruction spelled out in Eric. W. Sanderson's *Mannahatta: A Natural History of New York City* (New York: Abrams, 2016), 149.

3. "The Central Park," *New York Daily Times*, July 9, 1856.

4. Sara Cedar Miller, *Central Park, An American Masterpiece: A Comprehensive History of the Nation's First Urban Park* (New York: Abrams, 2003), 13, 242; "Statement of the Quantity of Certain Classes of Work Done and of Materials Used in the Construction of Central Park, Exclusive of Operations on the Central Water Works of the City, Third Annual Report," New York Department of Public Parks, 1874, 350–51.

5. The Romans invented this "tapping of springs" technique in ancient times.

6. Cited in: D. B. Stephens and D. A. Stephens, "British Land Drainers: Their Place among Pre-Darcy Forefathers of Applied Hydrogeology," *Hydrogeology Journal* 14 (2006): 1367–76 doi.org/10.1007/s10040-006-0052-1.

7. John H. Klippart, *The Principles and Practice of Land Drainage* (Cincinnati, OH: Robert Clarke & Co., 1867), 29.

8. Henry French, *Farm Drainage: The Principles, Processes, and Effects of Draining Land with Stones, Wood, Flows, and Open Ditches, and Especially with Tiles* (New York: Orange Judd & Co., 1859), 22.

9. George Perkins Marsh, *Man and Nature; or Physical Geography as Modified by Human Action* (e-book) (Project Gutenberg, 2011), 235, https://www.gutenberg.org/ebooks/37957.

10. Ibid., 35.

11. Henry David Thoreau, "Walking," 1862 essay reprinted in the *Atlantic*, November 2017.

12. Henry David Thoreau, "Walking, A Lecture," 1851, cited in: Sydney M. Williams, "Mud River Swamp," *Thought of the Day* (blog), May 24, 2017, https://swtotd.blogspot.com/2017/05/mud-river-swamp.html.

13. Solon Robinson, *Hot Corn: Life Scenes in New York Illustrated* (New York: De Witt and Davenport, 1854), 70.

14. Egbert L. Viele, "Central Park: Proposed Plan of Improvement" (report), *New York Daily Times*, February 20, 1857.

15. In both cases it was actually malaria.

16. Edwin Chadwick, "Metropolitan Sewage Committee Proceedings," *Parliamentary Papers 1846*, vol. 10, 651.

17. Quoted in: S. Halliday, "Death and Miasma in Victorian London: An Obstinate Belief," *BMJ* 323 (December 22, 2001): 1469–71, doi:10.1136/bmj.323.7327.1469.

18. "Poison: Gas Works" and "The Vampyre (No Superstition)," *Punch*, January–June 1846.]

19. *New England Farmer*, 1823, 326, https://books.google.ca/books?id=FbQVAAAAYAAJ&pg=PA326&lpg=PA326&dq=Old+Farmers+almanac+and+miasma&source=bl&ots=m_5T5FY8fs&sig=ACfU3U2gtSQzVJfpC3CrcUP3Xhk1zrBOzA&hl=en&sa=X&ved=2ahUKEwiwjtHa.

20. *Hostetter's United States Almanac, for the Use of Merchants, Mechanics, Farmers and Planters, and All Families*, 1864, https://www.abaa.org/book/1132023231.

21. Mary Wollstonecraft Shelley, *The Last Man*, Shelley, Mary (1826) (Whitefish, MT: Kessinger, 2004), 183.

22. Edgar Allan Poe, "The Fall of the House of Usher," in Tales and Sketches (London: Routledge, 1896), 129.

23. Harold Bloom, ed., *Modern Critical Views on Edgar Allan Poe* (New York: Chelsea House Publishers, 1985), 7.

24. Cited in: "Competing Theories of Cholera," UCLA Department of Public Health, Department of Epidemiology, n.d., https://www.ph.ucla.edu/epi/snow/choleratheories.html.

25. John E. Meyer, "William Farr on Cholera," *Journal of the History of Medicine* (April 1973): 190, http://www.medicine.mcgill.ca/epidemiology/hanley/temp/material/SnowCholera/EyleronWilliamFarrCholera.pdf.

26. "The New-York Sanitary Association.," *New York Times*, June 10, 1859.

27. *New England Farmer*, vol. 11, 1859.

28. Allen Boyer McDaniel, "Drainage of Farm Lands," *Bulletin of the University of South Dakota* 9, no. 6 (February 1910), https://babel.hathitrust.org/cgi/pt?id=loc.ark:/13960/t6tx3tw4v&view=1up&seq=3.

29. "Fuel: All about Peat," *New York Times*, October 19, 1865.

30. M. M. Weaver and Marion G. Leiby, "Farming History in Old Tile," *Soil*

Conservation 27–28 (August 1961): 250, https://books.google.ca/books?id=gAWd65BSseIC&pg=PA250&lpg=PA250&dq=Waterloo,New+York+and+tiles+factories&source=bl&ots=4QLaVIp6xg&sig=ACfU3U2lduenFe9Gd_wfrbhHovPcMFRV6Q&hl=en&sa=X&ved=2ahUKEwj42aCjucroAhUIv54KHeUFBQsQ6AEwAnoECAwQOA#v=onepage&q=Waterloo%2CNew%20York%20and%20tiles%20factories&f=false.

31. Anne Vileisis, *Discovering the Unknown Landscape: A History of America's Wetlands* (Washington, DC: Island Press, 1997).

32. See: Thomas J. Headlee, "The Mosquitoes of New Jersey and their Control," *New Jersey Agricultural Experiment Stations Bulletin* 348 (New Brunswick, NJ, 1921).

33. E. K. Soper and C. C. Osbon, "The Occurrence and Uses of Peat in the United States," *US Department of the Interior Bulletin* 728 (Washington, DC: US Geological Survey, 1922), https://pubs.usgs.gov/bul/0728/report.pdf.

34. William T. Hornaday to John K. Small, "Proposed Everglades National Park," December 30, 1932, Legislation, R. G. 79, File 120.

35. Thomas Dreschel et al., "Peat Soils of the Everglades of Florida USA," Peat, IntechOpen, 2017, open-access peer review chapter, https://www.intechopen.com/books/peat/peat-soils-of-the-everglades-of-florida-usa.

Chapter 3

1. William Byrd, *The Prose Works of William Byrd*, ed. Louis B. Wright (Cambridge, MA: Harvard University Press, 1966), 206, 240.

2. Virginia DeJohn Anderson, "Animals into the Wilderness: The Development of Livestock Husbandry in the Seventeenth-Century Chesapeake," *William and Mary Quarterly* 59, no. 2 (2002): 377–408, www.jstor.org/stable/3491742, accessed April 30, 2020.

3. Cited in: Roy T. Sawyer, *America's Wetland: An Environmental and Cultural History of Tidewater Virginia and North Carolina* (Charlottesville, VA: University of Virginia Press, 2010), 30.

4. Rufus A. Grider, "Early Navigation of the Mohawk," in *Papers Read Before the Herkimer County Historical Society During the Years 1896, 1897, and 1898* (repr., London: Forgotten Books, 2018), 111.

5. Anderson, "Animals into the Wilderness," 387.

6. Jared G. Beerman, "The Potential for Gray Wolves to Return to Pennsylvania Based on GIS Habitat Modeling," Department of Resource Analysis, Saint's Mary's University of Minnesota, Winona, MN.

7. Dave Foreman, *Rewilding North America: A Vision for Conservation in the 21st Century* (Washington, DC: Island Press, 2004), 71.

8. Michael K. Phillips, V. Gary Henry, and Brian T. Kelly, "Restoration of the Red Wolf," US Department of Agriculture: Animal and Plant Health Inspection Service, USDA National Wildlife Research Center—Staff Publications, University of Nebraska, January 2003, https://digitalcommons.unl.edu/icwdm_usdanwrc/234/.

9. William L. Hamnett and David C. Thornton, *Tar Heel Wildlife*, 2nd ed. (Raleigh, NC: North Carolina Wildlife Resources Commission, 1953), 48.

10. Antonio Rodriguez, Matthew Waters, and Milas Pehler, "Burning Peat and

Reworking Loess Contributes to the Formation and Evolution of a Large Carolina Bay Basin," *Quaternary Research* 77, no. 1 (January 2012).

11. Letter from Charles Darwin to Asa Gray in F. M. Jones, "'The Most Wonderful Plant in the World' with Some Unpublished Correspondence of Charles Darwin," *Natural History* 23 (1923): 589–96.

12. Cited in: Andrea Wulf, *The Brother Gardeners: Botany, Empire, and the Birth of an Obsession* (London: Windmill Books, 2009), 140.

13. Edmund Ruffin, *Agricultural, Geological, and Descriptive Sketches of Lower North Carolina, and the Similar Adjacent Lands* (Raleigh, NC: Institution for the Deaf & Dumb, & the Blind, 1861), https://lccn.loc.gov/04023143.

14. Quoted in: Roy T. Sawyer, *America's Wetland: An Environmental and Cultural History of Tidewater Virginia and North Carolina* (Charlottesville, VA: University of Virginia Press, 2010).

15. William Robbins, "Plantation is Carved Out of North Carolina Wilderness," *New York Times*, May 8, 1974.

16. Eric Carlson, "Stumpy Point Requests Runoff Regs," North Carolina Coastal Federation, *Coastal Review* 1, no. 3 (December 1983), 7.

17. "Peat Gasification Key to New Methanol Plant," *Chemical Engineering News* 59, no. 25 (June 22, 1981): 8–9, doi.org/10.1021/cen-v059n025.p008a.

18. US General Accounting Office, "Circumstances Surrounding the First Colony Peat-to-Methanol Project," Report to the Chairman, Subcommittee on Oversight and Investigations, Committee on Energy and Commerce, November 10, 1983, https://www.gao.gov/assets/150/140927.pdf.

19. Ibid.

20. Hervey Amsler Priddy, "United States Synthetic Fuels Corporation: Its Rise and Fall," (PhD dissertation, University of Texas at Austin, May 2013, 19.

21. Ibid.

22. Mary L. Duryea and P. M. Dougherty, *Forest Regeneration Manual* (Berlin: Springer Science, 1991), 404.

23. David Snyder, "Eggcentric: Farmer-Mogul Ruffles Feathers," *Crain's Chicago Business*, July 27, 1987, 1.

24. "Rose Acre Farms: Dynasty in Turmoil," *Indianapolis Star*, March 11, 1991.

25. Egg Industry Adjustment Act: Hearings, Ninety-second Congress, Senate, Committee on Agriculture and Forestry, Subcommittee on Agricultural Production, Marketing, and Stabilization of Prices, February 15 and 17, 1972, 63.

26. See, for example, three reports to the Atlanta district office of the US Department of Health and Human Services in 2018, citing Rose Acre Farms for unsanitary conditions: www.fda.gov/ucm/groups/fdagov-public/@fdagov-afda-orgs/documents/document/ucm604794.pdf.

27. "Marketing Savvy Grows Rose Acre Farms," *WATTPoultry*, October 3, 2008, https://www.wattagnet.com/articles/3269-marketing-savvy-grows-rose-acre-farms.

28. See the filing in the United States District Court for the Eastern District of North Carolina, Western Division, "Rose Acre Farms, Inc. v. North Carolina Department of Environment and Natural Resources," Case 5:14-cv-00147-D Document 25-1, filed May 14, 2014.

Chapter 4

1. S. B. Hodgkins et al, "Tropical Peatland Carbon Storage Linked to Global Latitudinal Trends in Peat Recalcitrance," *Natural Communications* 9, no. 3640 (2018), doi.org/10.1038/s41467-018-06050-2.

2. Monika K. Reczuga et al., "*Arcella peruviana sp. nov.* (Amoebozoa: Arcellinida, Arcellidae), a New Species from a Tropical Peatland in Amazonia," *European Journal of Protistology* 51, no. 5 (2015): 437–49.

3. P. K. L. Ng, J. B. Tay, and K. K. P. Lim, "Diversity and Conservation of Blackwater Fishes in Peninsular Malaysia, Particularly in the North Selangor Peat Swamp Forest," *Hydrobiologia* 285 (1994): 203–18, doi.org/10.1007/BF00005667.

4. Sri Suci Utami and Jan A. R. A. M. Van Hooff, "Meat-Eating by Adult Female Sumatran Orangutans (*Pongo pygmæus abelii*)," *American Journal of Primatology* 43, no. 2 (1997): 159–65.

5. L. S. Wijedasa et al., "Carbon Emissions from South-East Asian Peatlands Will Increase Despite Emission-Reduction Schemes," *Global Change Bio*logy 24 (2018): 4598–613, doi.org/10.1111/gcb.14340.

6. Hallett Hammatt, "Cultural Impact Assessment for the Alakai Protective Fence Project, Waimea and Wainiha Ahupua'a, Waimea and Hanalei Districts, Island of Kaua'i," report prepared for The Nature Conservancy, March 2008, http://hawp.org/_library/documents/news-and-announcements/eastalakaiculturalsurvey.pdf.

7. Alison R. Sherwood et al., "Freshwater Algae Associated with High-Elevation Bogs in the Hawaiian Islands," *Records of the Hawaii Biological Survey of 2012*, ed. Neal L. Evenhuis and Lucius G. Eldredge, *Bishop Museum Occasional Papers* 114 (2013): 21–31.

8. Lawrence Zettler and Steven Perlman, "*Peristylus holochila*—Hawaii's Rarest Native Orchid and the Battle to Save It from Extinction," *Orchids* 81 (2012): 94–99.

9. "Akikiki," *Field Guide to Birds of North America*, https://identify.whatbird.com/obj/1170/_/Akikiki.aspx.

10. Kathryn Hulme, "The Timeless Kaua'i Swamp," *Atlantic Monthly* 215, January 1965, 68–71.

11. Cited in: Fernando Penalosa, *The Alaka'i, Kauai's Unique Wilderness* (McKinleyville, CA: Quaking Aspen Books, 2010), 20–21.

12. Frederick B. Wichman, *Kauai: Ancient Place Names and Their Stories* (Honolulu, HI: University of Hawaii Press, 1998), 19.

13. Ibid., 14.

14. Frederick B. Wichman, *Na Pua Ali'i O Kaua'i: Ruling Chiefs of Kauai* (Honolulu, HI: University of Hawaii Press, 2003), 13.

15. Grant Harper and Nancy Bunbury, "Invasive Rats on Tropical Islands: Their Population Biology and Impacts on Native Species," *Global Ecology and Conservation* 3 (January 2015): 607–27.

16. Robert Wenkam, *Country Roads of Hawaii* (McKinleyville, CA: Country Roads Press, 1993).

Chapter 5

1. "Life Goes to the Death Valley Centennial," *Life* magazine, January 30, 1950.

2. Annual Report of the Secretary of the Department of the Interior, Fiscal Year Ending June 30, 1949, 335.

3. "A 100-pound male wolf was pursuing a bighorn sheep in the Mojave Desert's Providence Mountains in 1922 when a steel-jaw trap clamped onto one of its legs. In a 2015 report by scientist Bob Wayne, DNA evidence suggests it was a Mexican gray, the so-called lobo of the American Southwest." (Louis Sahagún, "DNA Indicates Long-Ago Southland Wolf Was Actually a Mexican Gray," *Los Angeles Times*, February 15, 2015.

4. Michelle Bushman, "Contribution of Recharge along Regional Flow Paths to Discharge at Ash Meadows, Nevada" (master's thesis, Brigham Young University, August 2008).

5. Randell J. Laczniak et al., "Estimates of Ground-Water Discharge as Determined from Measurements of Evapotranspiration, Ash Meadows Area," US Department of Energy Investigations Report: Issue 99, Part 4079, 1995–2000.

6. Teri A. Knight et al., "Status of Populations of the Endemic Plants of Ash Meadows, Nye Couty, Nevada," report to US Fish and Wildlife Service, Great Basin Complex, Reno, Nevada, Endangered Species Act—Section 6, Fiscal Year 1986, Project Agreement No. 86-2-1.

7. Stephen Nelson et al., "Regional Groundwater Flow in Structurally-Complex Extended Terranes: An Evaluation of the Sources of Discharge at Ash Meadows, Nevada," *Journal of Hydrology* 386, no. 1 (May 2010):118–29, doi:10.1016/j.jhydrol.2010.03.013.

8. Frank E. Rheindt et al., "A Lost World in Wallacea: Description of a Montane Archipelagic Avifauna," *Science 367, no. 6474 (*January 10, 2020): 167–70, doi:10.1126/science.aax2146.

9. Alan C. Riggs and James E. Deacon, "Connectivity in Desert Aquatic Ecosystems: The Devil's Hole Story," Conference Proceedings, Spring-fed Wetlands: Important Scientific and Cultural Resources of the Intermountain Region, Las Vegas, NV, May 5–7, 2002.

10. Leroy and Jean Johnson, eds., *Escape from Death Valley: As Told by William Lewis Manley and other '49ers* (Reno, NV: University of Nevada Press, 1987), 160.

11. Ibid., 55.

12. John Randolph Spears, *Illustrated Sketches of Death Valley and Other Borax Deserts of the Pacific Coast* (Chicago: Rand McNally, 1892).

13. Georgia Lewis, "Real Shotgun Kitty Bigger than Life," *Nevada West Pahrump Valley Times* (Pahrump, NV), April 3, 1973.

14. Ibid.

15. These so-called trains consisted of massive wagons hauling borax from the Harmony Borax Works near Furnace Creek to the railhead near Mojave, a grueling 165-mile, ten-day trip across primitive roads.

16. Eric J. Hilton and Gerald R. Smith, "The American Society of Ichthyologists and Herpetologists as an Advocacy Group: The Green River Poisoning of 1962," *Copeia* 1, no. 3 (December 2014): 577–91, doi.org/10.1643/OT-14-114.

17. Edwin Pister, "The Rare and Endangered Fishes of the Death Valley System, a Summary of the Proceedings of a Symposium Relating to their Preservation and Protection," Resources Agency of California, Department of Fish and Game, 1970.

18. Edwin Pister, "The Desert Fishes Council, Catalyst for Change," in *Battle*

Against Extinction: Native Fish Management in the American West, ed. W. L. Minckley and James E. Deacon (Tucson, AZ: University of Arizona Press, 1991).

19. Unsigned editorial in the *Elko (Nevada) Daily Free Press*, March 8, 1976; cited in W. L. Minckley and James E. Deacon, eds., *Battle Against Extinction: Native Fish Management in the American West* (Tucson, AZ: University of Arizona Press, 2017), 80.

20. US Department of the Interior, Fish and Wildlife Service, news release, July 21, 1968.

21. "Anaheim Brevities: Cattle Disease Abating. Value of Peat Land," *Los Angeles Times*, November 5, 1899.

22. "Orange County Brevities: Subterranean Lake Found on San Joaquin Ranch. Boundless in Extent and Unfathomed, It Underlies the Peat Lands—Clear Sulphur Water Unlike That of Surrounding Area," *Los Angeles Times*, July 11, 1897.

Chapter 6

1. Howard Crum, *A Focus on Peatlands and Mosses* (Ann Arbor, MI: University of Michigan Press, 1992).

2. Hanna R. Royals, Jean-François Landry, and Todd M. Gilligan, "The Myth of Monophagy in *Paralobesia* (Lepidoptera: Tortricidae)? A New Species feeding on *Cypripedium reginae* (Orchidaceae)," *Zootaxa* 4446, no. 1 (July 16, 2018): 81–96, doi:10.11646/zootaxa.4446.1.6.

3. Marko Mutanen et al., "*Monopis jussii*, a New Species (Lepidoptera, Tineidae) Inhabiting Nests of the Boreal Owl (*Aegolius funereus*)," *Zookeys* 992 (November 12, 2020): 157–81.

4. Graham A. Montgomery et al., "Standards and Best Practices for Monitoring and Benchmarking Insects," *Frontiers in Ecology and Evolution* 8 (2021): 513.

5. Kyle E. Johnson, "Natural History and Distribution of *Papaipema aweme* (Noctuidae)," *Journal of the Lepidopterists' Society* 71, no. 4 (201): 199–210.

6. David L. Wagner et al., "A Window to the World of Global Insect Declines: Moth Biodiversity Trends Are Complex and Heterogeneous," *Proceedings of the National Academy of Sciences* 118, no. 2 (Janurary 2021), doi:10.1073/pnas.2002549117.

7. Theodore Roethke, "Orchids," in *The Collected Poems of Theodore Roethke* (London: Faber & Faber, 1968).

8. O. A. Beath, J. H. Draize, and H. F. Eppson, "Arrow Grass—Chemical and Physiological Considerations," University of Wyoming Agricultural Experiment Station Bulletin 193, 1933.

9. Hans Joosten and J. Couwenberg, "Peatlands and Carbon," in: *Assessment on Peatlands, Biodiversity and Climate Change—Main Report* (Kuala Lumpur: Global Environment Centre, & Wageningen, Netherlands: Wetlands International, 2008), 99–117.

10. F. Riva, J. H. Acorn, and S. E. Nielsen, "Distribution of Cranberry Blue Butterflies (*Agriades optilete*) and Their Responses to Forest Disturbance from In Situ Oil Sands and Wildfires," *Diversity* 10 (2018), 112, doi.org/10.3390/d10040112.

11. Robin Wall Kimmerer, *Gathering Moss: A Natural and Cultural History of Mosses* (Corvallis, OR: Oregon State University Press, 2003).

Chapter 7

1. Seamus Heaney, "Kinship," from *North* (London: Faber & Faber, 1975).

2. Donald W. Buchanan, "James Wilson Morrice," *University of Toronto Quarterly* 2 (January 1937).

3. Paul H. Walton, "The Group of Seven and Northern Development," *RACAR: revue d'art canadienne / Canadian Art Review* 17, no. 2 (1990): 171–208, www.jstor.org/stable/42630464, accessed January 27, 2021.

4. Y. Uprety et al., "Traditional Use of Medicinal Plants in the Boreal Forest of Canada: Review and Perspectives," *Journal of Ethnobiology and Ethnomedicine* 8, no. 7 (2012): doi.org/10.1186/1746-4269-8-7.

5. Björn Hånell, Lars Lundin, and Tord Magnusson, "The Peat Resources in Sweden," in *After Wise Use—The Future of Peatlands, Proceedings of the 13th International Peat Congress: Peat In Energy* (Jyväskylä, Finland: International Peatland Society, 2008), https://peatlands.org/assets/uploads/2019/06/ipc2008p109-113-hanell-peat-resources-in-sweden.pdf.

6. Ibid.

7. Thanks to author Robert Macfarlane for making reference to this in his book *Landmarks* (New York: Penguin, 2016).

Chapter 8

1. Barbara Hurd, *Stirring the Mud: On Swamps, Bogs, and Human Imagination* (Athens, GA: University of Georgia Press, 2008).

2. B. Heidel and E. Rodemaker, *Inventory of Peatland Systems in the Beartooth Mountains, Shoshone National Forest, Park County, Wyoming*, report prepared for: Environmental Protection Agency, Wyoming Natural Diversity Database, Laramie, WY, 2008.

3. R. A. Chimner, J. M. Lemly, and D. J. Cooper, "Mountain Fen Distribution, Types and Restoration Priorities, San Juan Mountains, Colorado, USA," *Wetlands* 30 (2010): 763–71.

4. G. Hope, R. Nanson, and P. Jones, *Peat-Forming Bogs and Fens of the Snowy Mountains of NSW* (Sydney, Australia: Office of Environment and Heritage, 2012).

5. X. Pontevedra-Pombal et al., "Mountain Mires from Galicia (NW Spain)," in: *Peatlands: Evolution and Records of Environmental and Climatic Changes*, ed. I. P. Martini, A. M. Cortizas, and W. Chesworth (Amsterdam: Elsevier, 2006), 83–108.

6. Abraham Knechtel, "The Dominion Forest Reserves," Bulletin Number 3, (Canada) Department of the Interior, 1910, 26.

7. Ibid.

8. Walter Dwight Wilcox, *Camping in the Canadian Rockies: An Account of Camp Life in the Wilder Parts of the Canadian Rocky Mountains* (New York: Putnam, 1896), 143–44.

9. John Muir, "The Calypso Borealis," from *The Life and Letters of John Muir*, vol. 1, ed. William Frederic Badè (Boston and New York: Houghton. Mifflin, 1924).

Chapter 9

1. Amy Clampitt, *The Collected Poems of Amy Clampitt* (New York: Penguin Random House, 1999).

2. Terrence R. Fehner and F. Gosling, "Origins of the Nevada Test Site," History Division, Department of Energy, December 2000.

3. The Manhattan Project was a research and development undertaking during World War II that produced the first nuclear weapons. It was led by the United States with support of the United Kingdom and Canada.

4. John Clearwater and David O'Brien, "O Lucky Canada: Britain Considered Testing Nuclear Weapons in Northern Manitoba, but Found the Climate in Australia Much More Agreeable," *Bulletin of the Atomic Scientists* 59, no. 4 (2003), https://go.gale.com/ps/anonymous?id=GALE%7CA105163618&sid=googleScholar&v=2.1&it=r&linkaccess=abs&issn=00963402&p=AONE&sw=w.

5. Robert Bothwell, *Eldorado: Canada's National Uranium Company* (Toronto: University of Toronto Press, 1984), 92–116.

6. Evan Richardson, Ian Stirling, and David S. Hik, "Polar Bear (*Ursus maritimus*) Maternity Denning Habitat in Western Hudson Bay: A Bottom-up Approach to Resource Selection Functions," *Canadian Journal of Zoology* 83, no. 6 (June 2005): 860–70.

7. Clearwater and O'Brien, "O Lucky Canada."

8. Convention on the Conservation of European Wildlife and Natural Habitats, Standing Committee, 40th meeting, Council of Europe, Convention on the Conservation of European Wildlife and Natural Habitat, Strasbourg, December 1–4, 2020.

9. Elyn R. Humphreys et al., "Two Bogs in the Canadian Hudson Bay Lowlands and a Temperate Bog Reveal Similar Annual Net Ecosystem Exchange of CO2," *Arctic, Antarctic, and Alpine Research* 46, no. 1 (2014): 103–13, doi:10.1657/1938-4246.46.1.103.

10. C. D. Koven et al., "A Simplified, Data-Constrained Approach to Estimate the Permafrost Carbon–Climate Feedback," *Philosophical Transactions of the Royal Society A* 373, no. 2054 (November 13, 2015), doi.org/10.1098/rsta.2014.0423.

11. Gustaf Hugelius et al., "Large Stocks of Peatland Carbon and Nitrogen Are Vulnerable to Permafrost Thaw," *Proceedings of the National Academy of Sciences* 117, no. 34 (Aug 2020): 20438–46; doi:10.1073/pnas.1916387117.

12. Copernicus Atmosphere Monitoring Service (CAMS), "CAMS Monitors Unprecedented Wildfires in the Arctic," CAMS website, July 11, 2019, https://atmosphere.copernicus.eu/cams-monitors-unprecedented-wildfires-arctic.

13. Michelle C. Mack et al., "Carbon Loss from an Unprecedented Arctic Tundra Wildfire," *Nature* 475, no. 7357 (2011): 489, doi:10.1038/nature10283.

14. Evan Richardson, Ian Stirling, and Bob Kochtubajda, "The Effects of Forest Fires on Polar Bear Maternity Denning Habitat in Western Hudson Bay," *Polar Biology* 30 (2007): 369–78, doi:10.1007/s00300-006-0193-7.

Chapter 10

1. Clara Vyvyan, *The Ladies, the Gwich'in, and the Rat: Travels on the Athabasca, Mackenzie, Rat, Porcupine, and Yukon Rivers in 1926* (Edmonton, AB: University of Alberta Press, 1998).

2. National Snow and Ice Data Center, "All about Frozen Ground," NSIDC website, 2021, https://nsidc.org/cryosphere/frozenground/whereis_fg.html.

3. Gustaf Hugelius, "We Mapped the World's Frozen Peatlands—What We Found Was Worrying," *The Independent* (London), August 19, 2020, https://www.independent.co.uk/climate-change/news/peatland-natural-carbon-absorption-emission-climate-change-geoscience-a9671146.html.

4. In his book *Landmarks*, Robert Macfarlane does a masterful job of describing how concise Hebrideans were in describing every aspect and nuance of peat and the ecosystems that are associated with them.

5. Stuart Houston, ed., *Arctic Ordeal: The Journal of John Richardson, Surgeon-Naturalist with Franklin, 1820–1822* (Montreal: McGill–Queen's University Press, 1994), xxix.

6. Darwin Correspondence Project, "Letter no. 366F," https://www.darwinproject.ac.uk/letter/DCP-LETT-366F.xml, accessed March 1, 2020.

7. Not to be confused with *ugruk*, the word for bearded seal.

8. J. Ross Mackay, "Pingos of the Tuktoyaktuk Peninsula Area, Northwest Territories," *Géographie Physique et Quaternaire* 33, no. 1 (1979), https://www.erudit.org/en/journals/gpq/1979-v33-n1-gpq1495571/1000322ar.pdf.

9. "Great Reindeer Trek an Epic of the North," *New York Times*, May 14, 1933, https://timesmachine.nytimes.com/timesmachine/1933/05/14/105135949.html?pageNumber=145.

10. Stephen R. Brown, "The Great Canadian Reindeer Project," *Canada's History*, December 17, 2014, https://www.canadashistory.ca/explore/environment/the-great-canadian-reindeer-project.

11. Fred Bell, *Geological Hazards: Their Assessment, Avoidance and Mitigation* (Boca Raton FL: CRC Press, 1999), 616.

12. Canadian Environmental Assessment Agency, "Oil and Gas Development Support Operations at Camp Farewell in the Kendall Island Bird Sanctuary," Canadian Environmental Assessment Registry: 05-01-14238, December 5, 2012, https://iaac-aeic.gc.ca/052/details-eng.cfm?pid=14238.

13. Environment Canada, "Mackenzie Gas Project Environmental Assessment Review Written Submission," Topic 7: Wildlife and Wildlife Habitat Migratory Birds including Kendall Island Bird Sanctuary, November 15–16, 2006, https://docs2.cer-rec.gc.ca/ll-eng/llisapi.dll/fetch/2000/90464/90550/338535/338661/343078/346094/603301/ENVCA%2D20F__%2D_Exhibit_5%2D_Environment_Canada_Mackenzie_Gas_Project_Environmental_Assessment_Written_Submission_Theme_3_and_Topic_7%2C_November_2006_%28J%2DEC%2D00136%29_%28A1S0R9%29.pdf?nodeid=603504&vernum=-2.

14. Amber R. Ashenhurst and Susan J. Hannon, "Effects of Seismic Lines on the Abundance of Breeding Birds in the Kendall Island Bird Sanctuary, Northwest Territories, Canada," *Arctic* 61, no. 2 (June 2008): 190–98.

15. Hamlet of Tuktoyaktuk, Town of Inuvik, Government of Northwest Territories, "Environmental Impact Statement for Construction of the Inuvik to Tuktoyaktuk Highway, NWT," Appendix F, "Inuvik to Tuktoyaktuk: All-Weather Road Economic Analysis," EBA File: V23201322.006, May 2011, https://www.inf.gov.nt.ca/sites/inf/files/resources/ith_eis.pdf.

16. Eric Bowling, "Great Slave MLA Katrina Nokleby Calls for Acceleration of

Inuvik–Tuktoyaktuk Highway Repairs," *NNSL Media*, November 8, 2020, https://www.nnsl.com/inuvikdrum/great-slave-mla-katrina-nokleby-calls-for-acceleration-of-inuvik-tuktoyaktuk-highway-repairs/.

17. ConocoPhillips Canada Resources Corporation and Shell Canada Limited, 2019 ABQB 727 (CanLII), Court of Queen's Bench of Alberta, https://canlii.ca/t/j2gwn, accessed January 30, 2021.

18. D. M. Masterson, "The Arctic Islands Adventure and Panarctic Oils Ltd.," *Cold Regions Science and Technology* 85 (2013): 1–14.

19. Judge Michel Bourassa, Provincial Court of the Northwest Territories, "The Role of the Provincial Courts and Criminal/Quasi Criminal Justice Process," originally presented at the conference, Law and the Environment: Problems of Risk and Uncertainty, Dalhousie University, Halifax, Nova Scotia, October 15, 1988.

20. H. M. French, "Surface Disposal of Waste-Drilling Fluids, Ellef Ringnes Island, N.W.T.: Short-Term Observations," *Arctic* 38, no. 4 (December 1985): 292–302.

Chapter 11

1. Fisheries and Oceans Canada, Central and Arctic region, "Babbage River Dolly Varden," Stock Status Report D5-62, https://waves-vagues.dfo-mpo.gc.ca/Library/316822.pdf.

2. T. C. Lantz and K. W. Turner, "Changes in Lake Area in Response to Thermokarst Processes and Climate in Old Crow Flats, Yukon," *Journal of Geophysical Research: Biogeosciences* 120, no. 3 (March 2015): 513–24, doi:10.1002/2014JG002744.

3. National Snow and Ice Data Center, "Methane and Frozen Ground," *All About Frozen Ground* (NSIDC website), 2021, https://nsidc.org/cryosphere/frozenground/methane.html.

4. Merritt R. Turetsky et al., "Permafrost Collapse Is Accelerating Carbon Release," *Nature* 569 (May 2, 2019).

5. S. V. Kokelj, B. Zajdlik, and M. S. Thompson, "The Impacts of Thawing Permafrost on the Chemistry of Lakes across the Subarctic Boreal Tundra Transition, Mackenzie Delta Region, Canada," *Permafrost and Periglacial Processes* 20, no. 2 (April/June 2009): 185–99, doi.org/10.1002/ppp.641.

6. Karen Graham, "Warming Temperatures Cause Canadian Lake to Fall Off Cliff," *Digital Journal*, December 21, 2015, http://www.digitaljournal.com/news/environment/dramatic-effect-of-climate-change-canadian-lake-falls-off-cliff/article/452764.

7. P. F. Schuster et al., "Permafrost Stores a Globally Significant Amount of Mercury, *Geophysical Research Letters* 45, no. 3 (2018): 1463–71, doi.org/10.1002/2017GL075571.

8. Trevor Lantz et al., "Warming-Induced Shrub Expansion and Lichen Decline in the Western Canadian Arctic," *Ecosystems* 17, no. 7 (November 2014): 1151–68.

9. A. Edwards et al., "A Distinctive Fungal Community Inhabiting Cryoconite Holes on Glaciers in Svalbard," *Fungal Ecology* 6, no. 2 (2013):168–76, doi.org/10.1016/j.funeco.2012.11.001.

10. Barun Majumder, "Wind Model Ideal for Drift Formation on Gentle Slopes," Cold Region Research Center Day, Wilfrid Laurier University, November 29, 2018, http://sdw.enr.gov.nt.ca/nwtdp_upload/11-Majumder-CRRC-Nov2018-CIMP200.pdf.

Chapter 12

1. Emily Dickinson, "Sweet Is the Swamp with Its Secrets," *The Complete Poems of Emily Dickinson* (Boston: Little, Brown, 1924).

2. Tessa Plint, Fred Longstaffe, and Grant Zazula, "Giant Beaver Palaeoecology Inferred from Stable Isotopes," article number 7179, *Scientific Reports* 9 (May 2019).

3. Jarrett A. Lobell, "Oldest Bog Body," *Archaeology*, January/February 2014.

4. Richard C. Hulbert Jr. and C. Richard Harington, "An Early Pliocene Hipparionine Horse from the Canadian Arctic," *Paleontology* 42, no. 6 (December 1999): 1017–25, doi.org/10.1111/1475-4983.00108.

Chapter 13

1. Tim Robinson, *Connemara: Listening to the Wind* (London: Penguin, 2008).

2. N. Roberts et al., "Europe's Lost Forests: A Pollen-Based Synthesis for the Last 11,000 Years," *Scientific Reports* 8, article no. 716 (January 2018), https://www.nature.com/articles/s41598-017-18646-7.

3. Petra J. E. M. van Dam, "Sinking Peat Bogs: Environmental Change Holland, 1350–1550," *Environmental History* 6, no. 1 (2001): 32–45.

4. Robert Angus Smith, "A Study of Peat," *Manchester Literary and Philosophical Society Proceedings* (1876): 156–57.

5. W. H. Wheeler, "The Commercial Uses of Peat," *Nature* 63 (1901): 590–91.

6. Lee F. Klinger, "The Myth of the Classic Hydrosere Model of Bog Succession," *Arctic and Alpine Research* 28, no. 1 (1996): 1–9, doi:10.1080/00040851.1996.12003142.

7. Alan Mairson, "The Everglades: Dying for Help," *National Geographic* 185, no. 4 (April 1994): 2–35.

8. Olena Volik et al., "Wetlands in the Athabasca Oil Sands Region: The Nexus between Wetland Hydrological Function and Resource Extraction," *Environmental Views* 28, no. 3 (2020).

9. R. C. Rooney, S. E. Bayley, and D. W. Schindler, "Oil Sands Mining and Reclamation Cause Massive Loss of Peatland and Stored Carbon," *Proceedings of the National Academy of Sciences* 109, no. 13 (2012): 4933–37.

Conclusion

1. United States Army Corps of Engineers v. Hawkes Co., Inc., et al., U.S. 15-290 (2016), https://www.supremecourt.gov/opinions/15pdf/15-290_6k37.pdf.

2. Barb Ganter and Hugh Boyd, "A Tropical Volcano, High Predation Pressure, and Breeding Biology of Arctic Birds: A Circumpolar Review of Breeding Failure in the Summer of 1992," *Arctic* 53, no. 3 (September 2000), doi.org/10.14430/arctic859.

3. Megha Verma et al., "Can Reindeer Husbandry Management Slow Down the Shrubification of the Arctic?" *Journal of Environmental Management* 267, no. 110636 (2020), doi:10.1016/j.jenvman.2020.110636.

4. Evan Reimondo, Thomas D Sisk, and Tad Theimer, "Effects of Introduced Bison on Wetlands of the Kaibab Plateau, Arizona," in *The Colorado Plateau VI: Science and Management at the Landscape Scale* (Tucson, AZ: University of Arizona Press, Jan 2015).

5. John Schwartz, "Big Step Forward for $50 Billion Plan to Save Louisiana Coast," *New York Times*, March 5, 2021.

6. Merritt Turetsky, presentation made to the Permafrost Carbon Feedback Action Group symposium, cited in "Permafrost Carbon Feedback Is Reducing the Opportunity to Avoid Global Climate Crisis," March 2021, https://cascadeinstitute.org/wp-content/uploads/2021/03/PCF-Dialogue-3-Summary-Report-Where-to-From-Here-March-2021.pdf.

7. "Legal Challenge over River Tay's Wild Beavers," *BBC News*, March 2, 2011.

8. Alan Law et al., "Using Ecosystem Engineers as Tools in Habitat Restoration and Rewilding: Beaver and Wetlands," *Science of the Total Environment* 605–6 (2017): 1021–30.

9. Steven M. Fountain, "Ranchers' Friend and Farmers' Foe: Reshaping Nature with Beaver Reintroduction in California," *Environmental History* 19, no. 2 (April 2014): 239–69.

10. Susan Charnley, "Beavers, Landowners, and Watershed Restoration: Experimenting with Beaver Dam Analogues in the Scott River Basin, California," Research Paper PNW-RP-613, US Department of Agriculture, US Forest Service, 2018, doi.org/10.2737/PNW-RP-613.

11. A. Huertas Herrera et al., "Mapping the Status of the North American Beaver Invasion in the Tierra del Fuego Archipelago," *PLOS ONE* 15, no. 4 (2020), doi.org/10.1371/journal.pone.0232057.

12. J. Leifeld and L. Menichetti, "The Underappreciated Potential of Peatlands in Global Climate Change Mitigation Strategies, *Nature Communications* 9, article no. 1071 (2018), doi.org/10.1038/s41467-018-03406-6.

13. John Cannon, "Oil Exploration at Odds with Peatland Protection in the Congo Basin," *Mongabay*, March 23, 2020, https://news.mongabay.com/2020/03/oil-exploration-at-odds-with-peatland-protection-in-the-congo-basin/.

14. F. Riva and S. E. Nielsen, "A Functional Perspective on the Analysis of Land Use and Land Cover Data in Ecology," *Ambio* 50 (2021): 1089–1100, doi.org/10.1007/s13280-020-01434-5.

15. Cassidy K. van Rensen et al., "Natural Regeneration of Forest Vegetation on Legacy Seismic Lines in Boreal Habitats in Alberta's Oil Sands Region," *Biological Conservation* 184 (201): 127–35.

16. M. Strack et al., "Petroleum Exploration Increases Methane Emissions from Northern Peatlands, *Nature Communications* 10, article no. 2804 (2019): doi.org/10.1038/s41467-019-10762-4.

17. "Doubling Back and Doubling Down: G20 Scorecard on Fossil Fuel Funding," *Africa Energy Portal* (website), November 12, 2020, https://africa-energy-portal.org/reports/doubling-back-and-doubling-down-g20-scorecard-fossil-fuel-funding.

Acknowledgments

One thing that became clear while I was writing this book is the connection between peatlands and indigenous people. Pristine peatlands like those in the Hudson Bay Lowlands, the Ramparts Wetlands in the Northwest Territories, and the Old Crow Flats in the Yukon are the homelands and hunting grounds to the Cree, the Sahtú, and the Gwitch'in, respectively. The Tlingit, Haida, and Tsimshian Tribes have just as intimate a relationship with the Tongass National Forest in Alaska as do the Chipewyan, Cree, and Métis with the fens and bogs of northern Alberta and northern Saskatchewan. The Hudson Bay Lowlands in northern Manitoba, Ontario, and Quebec are home to the Mushkegowuk First Nations. The Omushkegowuk have navigated the rivers, wetlands, and marine regions there since long before Europeans arrived on the scene.

Large peat-accumulating wetlands, termed *bofedales* in the Andes, are critical pastures for indigenous people like the Aymaran highlanders who live in the *puna* of Bolivia and the Tibetans who herd yaks and other animals on the Zoigê Plateau.

Many of these indigenous people fear that further incursions into peatlands by miners, foresters, oil and gas extractors, and hydroelectric developers will spell the end of their way of life. It's happened all too often in the past. The Nansemond no longer dwell in the Great Dismal Swamp or anywhere else on the Albermarle Peninsula. The Ottawa are gone from the Black Swamp of Ohio and neighboring states. The Wintun, Maidu, Miwok, and Yokut tribes no longer have a place in the Sacramento–San Joaquin River Delta and other peatlands of California. The Southern Paiute and Timbisha Shoshone finally do have some say in how the Ash Meadows National Wildlife Refuge is managed in the Mojave Desert, but there is not much left in the way of peatland to manage because of climate change and human disturbances.

An acknowledgment of the highest order goes to them and many other indigenous people.

I'd like to thank the many scientists and national refuge managers for inviting me to join them in the field: John Acorn, Andrew Derocher, Mike Demuth, John England, Bert Finnamore, Darryl Hedman, Sandrine Hugron, Kyle Johnson, Randy Kolka, Corey Lee, Phil Marsh, Howard Phillips, Greg

Pohl, John Pomeroy, Line Rochefort, Eric Soderholm, Mike Waddington, Fred Wurster, and many others I talked to on the phone and in e-mails. Thanks as well to the National Tropical Botanical Gardens in Hawaii for offering me an internship that put me in the field with botanists Steve Perlman and Ken Wood, who have done more for plant conservation in the South Pacific than anyone else. And I cannot forget ranchers Mac Blades and Fran Gilmar in Alberta for giving me insights into fens and wetlands on their properties.

A special thanks to peatland ecologists Dale Vitt and David Cooper for being so helpful and taking the time to point out things I did not get quite right and other things that I might have otherwise missed. There are few scientists in this field who are as seasoned and accomplished as these two. And thanks as well to scientist Duane Froese who invited me to listen in on four insightful scientific brainstorming sessions on permafrost, the role it plays in climate change and the carbon cycle, and the ways in which it might be managed.

This is my third book with Island Press editor Courtney Lix. I'd like to say that our relationship has gotten better with each book, but because it was always as good as it gets, I can only thank her once again for her wisdom, advice, and encouragement. I say the same about my agent, Lisa Adams, who spent an inordinate amount of time getting me to focus on the outline before she offered it up to Island Press. I know that many of my colleagues are jealous that I have an agent as diligent and demanding as Lisa is.

Michael Fleming is a creative editor, novelist, poet, and teacher of composition and literature. That's all I need to say to explain why I was so humbled by his superb copyediting. None of this, of course, could have been done without the keen oversight of Sharis Simonian, the production manager at Island Press.

Finally, hats off once again to my wife, Julia, for joining me on several of these expeditions and for listening to my story ideas and helping me get through the writing process, which, for the first time, did not include our children because they are now living in other cities and leading their own adventurous lives. Last but not least, a thumbs-up to my new office partner, Saska, an unspeakably sweet golden retriever pup, who reminded me three times daily that I needed to take a break and go for a walk to clear my mind.

Index

Page references in *italics* indicate photographs.

About the Author

Edward Struzik has been writing about scientific and environmental issues for more than thirty-five years. A fellow at the Institute for Energy and Environmental Policy in the School of Policy Studies at Queen's University in Kingston, Canada, his numerous accolades include the Atkinson Fellowship in Public Policy and the Sir Sandford Fleming Medal, awarded for outstanding contributions to the understanding of science. In 1996 he was awarded the Knight Science Journalism Fellowship and spent a year at Harvard and MIT researching issues of the environment, evolutionary biology, and politics. He has been a regular contributor to *Yale Environment 360*, an online magazine published by the Yale School of Environment, since it was launched in 2008.

Island Press | Board of Directors